C0-DYB-706

Making a Disney WISH

Behind the Dreams of Disney's Newest Ship

David J. Fisher   Foreword by Thomas Mazloum

Disney EDITIONS
Los Angeles · New York

Copyright © 2022 Disney Enterprises, Inc.

All rights reserved. Published by Disney Editions, an imprint of Buena Vista Books, Inc.
No part of this book may be reproduced or transmitted in any form or by any means,
electronic or mechanical, including photocopying, recording, or by any information
storage and retrieval system, without written permission from the publisher.

For information address Disney Editions, 77 West 66th Street, New York, New York 10023.

ISBN: 978-1-368-07777-4
FAC-034274-22133

Printed in the United States of America
First Hard Cover Edition, June 2022
10 9 8 7 6 5 4 3 2 1
Visit www.disneybooks.com

Contents

Deck 1 — When You Wish Upon a *Wish* 1

Deck 2 — It All Started with a Breakdown 11

Deck 3 — If It Ain't Broke, Fix It Anyway 19

Deck 4 — The Theme of It All 31

Deck 5 — Stories and Characters and Motifs, Oh My! 43

Deck 6 — It Takes a (Worldwide) Village to Raise a Ship 51

Deck 7 — Lockdown, *Schmockdown!* 61

Deck 8 — Full Speed Ahead 71

Deck 9 — A Hall So Grand That's Exactly What We Named It 87

Deck 10 — What to Expect When You're Exploring 97

Deck 11 — Eat, Drink, and See Elsa 111

Deck 12 — All the Ship's a Stage (But Mostly the Walt Disney Theatre) 121

Deck 13 — It's a Bird! It's a Plane! It's AquaMouse! 131

Deck 14 — The Whole *Youth* and Nothing but the *Youth* 139

Deck 15 — The State of the Rooms 149

Deck 16 — A Dream Is a Wish Your *Art* Makes 161

Deck 17 — Wish Upon a Star and Save the Universe 169

Deck 18 — A *Wish* Fulfilled 177

Foreword

The story of the *Disney Wish* is the story of imagination. It is the story of bold new ideas and how a team can build something so distinctively *Disney* that it surprises everyone. Most of all, it is the story of a group of highly creative and forward-thinking individuals, including Imagineers, designers, producers, and cast and crew, all of whom are relentless in their pursuit of excellence.

This book takes us on a journey into the very essence of our newest cruise ship. It offers telling insights into the design and construction process from the visionaries and builders themselves, and it reveals details about every nook and cranny of the vessel. It is both a front-row seat and a behind-the-scenes study of how a world-class Disney ship is brought to life.

When I began my Disney career back in 1998, I had much to learn about Disney Cruise Line and its unique position in the industry. Though I was no stranger to cruising and hospitality, it was immediately apparent to me that Disney was different, focused on creating truly personalized experiences, all rooted in beloved Disney story lines. As a hotel director on board the *Disney Magic*, I quickly adopted what it meant to deliver a "magical moment" and to go from "good to great" whenever possible. Since those early days, serving as a senior leader at Walt Disney World Resort and at Disney Signature Experiences, I have never lost sight of what makes Disney so enchanting, whether it's attentive guest service or a touching stage show that turns the audience joyously misty-eyed.

With the *Disney Wish*, we are navigating to new horizons while staying true to our fundamentals. Think of it this way: this ship gives families exactly what they expect from a Disney cruise vacation while exceeding their expectations at every opportunity. Sounds fun, doesn't it? I encourage you to enjoy the stories and anecdotes in this volume shared directly from the people responsible for envisioning and creating the *Disney Wish*.

You will understand the passion they brought to their work and the inspiration they felt in bringing forth something so incredible, something that will dazzle families for many years to come. Claire Weiss, an interior designer with Walt Disney Imagineering, said it well: "I don't think you ever get used to the scale and magnitude of this project, and I don't think we even completely wrap our minds around it. It really is this sense of specialness and this feeling that, wow, this thing we're working on truly is incredible."

Incredible indeed. From its jaw-droppingly gorgeous Grand Hall to the stunningly ornate Walt Disney Theatre, and from the whimsical AquaMouse (Disney's first-ever attraction at sea) to the interactive Disney Uncharted Adventure that uses augmented reality and physical effects to tell Disney stories in completely new ways—the *Disney Wish* provides so much more than a cruise. It is, as we might say, a dream come true.

As we welcome our fifth ship to the Disney Cruise Line fleet, I want to thank the diverse and talented team members who not only made this enormous project possible but pushed themselves and raised the bar for innovation every step of the way. Their figurative fingerprints are on every aspect of the ship and experience, from stem to stern. And now, still awash in pixie dust, we are wasting no time in preparing for our next two ships. You can bet that our fresh ideas and "break the mold" thinking will inform our progress as we work to make sure everyone who sails with us takes away a lifetime of cherished memories.

Anchors away!

Thomas Mazloum

President,
Disney Signature Experiences

After more than five years of planning, design, and construction, the *Disney Wish* set sail on its maiden voyage in summer 2022.

When You Wish Upon a *Wish*

When the *Disney Wish* was floated out of Hall 6 at the Meyer Werft shipyard in early February—bringing forth feelings of warmth and joy to a typically cold winter's day—in Papenburg, Germany, the sparkling new ship was greeted with all the hoopla and fanfare it deserved: a rousing musical overture, celebratory fireworks, tossed fluttering balls of confetti, shouts of "Huzzah!" . . . and, from those who had brought it to life, huge sighs of relief.

The *Disney Wish* was not an easy ship to birth (pardon the pun), not that any project of this size and scope is without its trials, tribulations, and travails. There are certainly going to be design challenges, production snags, and construction difficulties that will always have to be confronted and overcome on any project.

But this one was designed and partially built in the most unusual of circumstances: during a worldwide pandemic lockdown that not only forced team members to work out of their kitchens and living rooms for well over a year but required health and safety protocols at the warehouses, mock-up spaces, and shipyards where it was being built and assembled.

And yet, as the ship in all its glory was pulled out of that massive hall fully intact, the team knew that it was worth every exhausting, exasperating minute it took to design and build it! They lived for this moment, when it all became tangible, actual, physical. There are those who will tell you that no Disney cruise ship is "better" than the other Disney cruise ships. And that's undeniably true—they're all unique and distinct and remarkable. But let's face it, on this day, this newest one . . . it's real and it's spectacular.

The *Disney Wish* is similar to the other Disney cruise ships, enhanced with lots of new features.

"The rotational restaurants, the live entertainment venues, it was all very innovative at the time on those first ships and are still extremely popular with our guests. We've just been able to plus that up with every ship we've built," said Bob Tracht, design and program management executive, Walt Disney Imagineering. (Tracht was one of the first people to join the design team for the *Disney Magic* and has worked on every Disney cruise ship since.)

"For instance, on the *Disney Dream* and the *Disney Fantasy*, we were able to make a larger cinema and a larger Walt Disney Theatre," noted Tracht. "Plus we were able to take advantage of advanced technology."

"And now on the *Wish*," he added, "we're just taking it to a whole different level."

"The *Disney Wish* is truly a whole new ship," said Portfolio Creative Executive Laura Cabo, the Imagineer who led the creative design effort for Disney Cruise Line's latest spectacular. "Not only in what we are offering from an experience and design point-of-view, but also from the fundamental layout of the ship."

The first thing one will undoubtedly notice upon stepping aboard *Wish* is the interior. It's different and in a big way . . . literally! The Grand Hall (the central meeting space, which is called the Atrium on the other Disney cruise ships) is massive, permitting not only more space and better guest flow, but featuring a magnificent chandelier as well, a signature statue, and a stage with accompanying balcony created especially for Disney character performances and appearances.

There are new, expanded locations for the popular, adult-oriented Senses Spa and the youth-driven Disney's Oceaneer Club; plus several new venues have been added. These include a double-height multipurpose venue called Luna, and the Hero Zone, an indoor events space. And then there's AquaMouse, Disney's first attraction-at-sea.

Another big change has also taken place "under the hood," so to speak. Among the many enhancements that will make the *Disney Wish* one of the safest, most efficient, and environmentally friendly ships at sea is that it's one of the first vessels to be powered by liquefied natural gas (LNG), a cleaner burning fuel than diesel, the current industry standard.

The largest transformation, though, is probably the most subtle: the introduction of an underlying and unifying motif that ties every aspect of the ship together on both a thematic and emotional level. There's a reason this ship is called the *Disney Wish*—and it has to do with something that is at the core of so many Disney stories: *enchantment*.

"Enchantment is about unlocking the wonder in everything we do and experience," said Cabo. "It has allowed us to create a signature design language that infuses every

As seen here in an artist's rendering, Rapunzel (with Pascal) became the latest Disney character—and the first princess—to grace the stern of a Disney cruise ship.

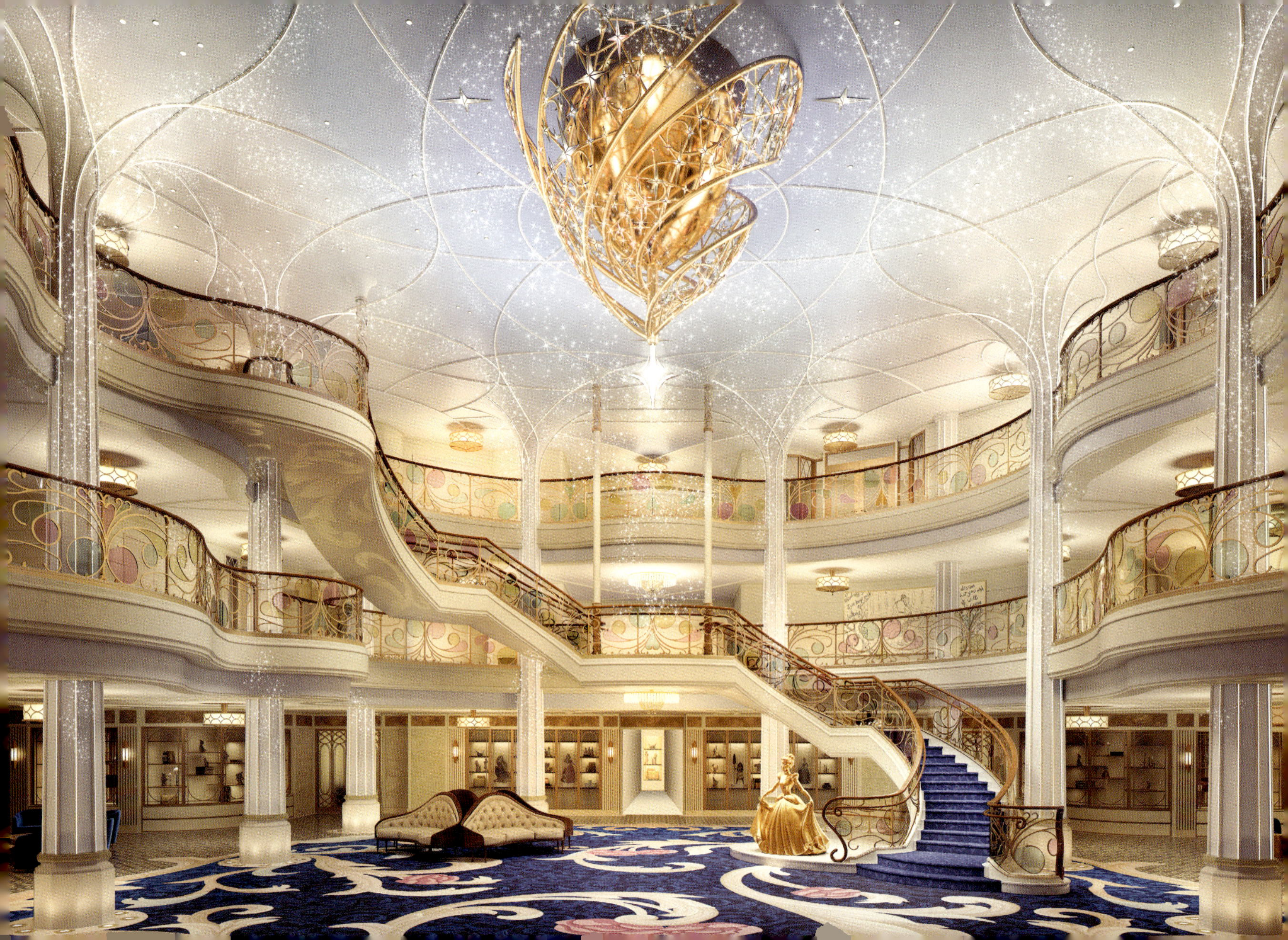

The Grand Hall (left), with Cinderella and a massive light fixture at its center, is the most expansive central gathering space ever created for a Disney cruise ship.

Among the many technological advancements aboard the *Disney Wish* are the propellers (top). Known as azimuth thruster pods, they make it easier to maneuver the ship, especially when it's entering and leaving ports.

The ship's final construction block, consisting of the Quiet Cove Pool (bottom left), the premier dining restaurants Enchanté and PALO Steakhouse, and Marceline Market, is hoisted into place on August 27, 2021.

A commemorative coin was placed in the keel to celebrate the beginning of the ship's construction (bottom right). The *Disney Wish* features Captain Minnie Mouse.

The *Disney Wish* is everything guests have come to expect from a Disney cruise ship—and more. Among the new to the Disney Cruise Line offerings are *Star Wars*: Hyperspace Lounge (top), where one can enjoy a cocktail "a long time ago in a galaxy far, far away"; Luna (center left), a two-deck-high entertainment space that hosts everything from family-friendly games and experiences to nighttime adults-only activities; an outdoor relaxation space at Senses Spa (center right), located in the ship's bow; Worlds of Marvel (bottom left), a restaurant that combines dinner with a demonstration of quantum technology by several Avengers; and Hero Zone, an indoor multipurpose space for games and other physical activities.

inch of the ship with more extensive Disney storytelling. We're telling princess stories, Marvel stories, *Star Wars*, pirates; and they're all threaded together through this theme."

"Taking a Disney cruise is a big milestone for a lot of families," said Denise Case, entertainment creative director, Disney Cruise Line. To bring families together, to have them enjoy each other, enjoy their favorite stories, their favorite characters all together, we know that that is so precious. And we just want to make sure we deliver the best guest experience ever."

And that's what the *Disney Wish* does. But the thing about Disney Cruise Line ships is that, although they may seem effortlessly enchanting and infinitely entrancing, the way they are designed and built requires a lot more than just wishing and dreaming. They certainly don't just materialize with a wave of a Fairy Godmother's magical wand.

This massive project took the imagination, ingenuity, creativity, organization, hard work of hundreds of Imagineers and their partners at Disney Cruise Line (DCL). They brainstormed and debated (and hemmed and hawed) over what to do differently with the *Disney Wish* (and her two sister ships that will follow in 2024 and 2025) for almost two years before locking in the layout, makeup, design, and—perhaps mostly importantly—spirit of the ship that proudly sails through the Caribbean and around the Bahamas today.

Along with collaborators at Meyer Werft, as well as an array of design firms, outfitters, contractors, and consultants from around the world, Imagineers never lost sight of who they were developing and designing a cruise ship for.

This is their story, though there is a spoiler alert: it *will* end happily ever after.

Bob Tracht

Executive Cruise Ship Design and Program Management Director, Walt Disney Imagineering

Bob Tracht has been with the Walt Disney World Resort since its very beginning. Actually, since before it began.

"My dad was involved in the Chamber of Commerce in Lima, Ohio," said Bob, who grew up in Cambridge, "and he was invited to come to Walt Disney World to be part of the pre-opening test group. He took me with him, so I got to be at Magic Kingdom while they were filming the grand opening event. I was five."

At sixteen, he returned as a videographer for a high school band. Never mind that it was the band of a rival high school—he got to go Walt Disney World again.

"On the way down, the video machine broke," he said. When I got to Florida, there was no place to get it repaired. It was the weekend, so there was no place to rent one. I get to Disney and the person who is responsible for the Magic Music Days program took me down into the tunnels beneath the Magic Kingdom and within twenty minutes he had the exact same machine that I needed.

"Not only that, he paired me with a media person and we spent two-and-a-half days documenting the band's experience. At that point, I said to myself, 'If a company can do this in twenty minutes, I want to work for that company.' And that's when I decided I wanted to work for Disney."

And really, it wasn't that long before Bob was hired by Disney. Just four years later, he was working as an audio technician on the shows that were performed in front of Cinderella Castle in the Magic Kingdom.

Shortly after that, he joined Walt Disney World Creative Entertainment, handling the technical elements for live Disney shows around the world and then helping start entertainment operations for the resort's new convention business. It was also at this time that he had his first exposure to Walt Disney Imagineering, working as a liaison on such projects as Disney's BoardWalk Inn and Disney's Animal Kingdom.

That relationship led to his becoming an Imagineering and, perhaps, his biggest and best assignment yet.

"I was one of the first sixteen people in the cruise company," he said. "I was brought on even before the president of the cruise line.

"I was one of the only people on the project who had experience with attractions and the more traditional things that Imagineering would build," he continued, "so I was really the voice of, 'Hey we have to hide all this technology. Hey, we have to do this.' I was just taking all the things I had learned on land and applying them to this thing that moved, in water."

Bob is one of the few people who has worked on every Disney cruise ship.

"I've never had a job," he said. "It's always been a hobby."

The upper decks of the *Disney Wish* feature an adults-only Quiet Cove area that offers stunning views off the back of the ship and, winding its way above and around six mid-ship pools, Disney's first-ever attraction at sea, AquaMouse.

Sharon, Lilly, Diane, and Walt on deck of *RMS Queen Elizabeth*, 1949, while Walt and family visited Europe during Walt's production set visit for *Treasure Island*.

It All Started with a Breakdown

Marty Sklar was asked the same question whenever Disney did something new and different, knowing full well that the interviewer was hinting that the company's founder might not approve: "What would Walt say?" The late Disney Legend, who worked side by side with Walt Disney in the early 1960s and eventually became the creative leader of Walt Disney Imagineering for more than thirty years, was always ready with the same reply: "Walt would say, 'What took you so long?'"

That was no doubt true about Disney's entry into the cruise business in 1998. Yes, it was a welcome development for a company that had long been at the forefront of family entertainment experiences, but it was also long overdue. Not that The Walt Disney Company hadn't been exploring the possibility for many years prior to that.

In fact, an argument can be made that Walt Disney himself probably would have steered his company into the cruise ship business had he lived longer (or had more money to invest in it earlier in his life). Walt loved cruising, dating back to 1931 when he discovered its benefits after suffering through one of the lowest points of his life.

Still struggling to make his studio successful, even after the breakthrough success of Mickey Mouse three years earlier, Walt Disney worked himself to exhaustion. On doctor's orders, he left the studio in the care of his brother Roy O., while he and his wife, Lillian, took their first vacation since their honeymoon five years earlier.

Part of their itinerary included a cruise through the Panama Canal, from Havana, Cuba, to Los Angeles. And according to Bob Thomas, in his book *Walt Disney: An American Original*, "Walt and Lilly agreed that was the best part of the trip."

Walt and Lillian (and sometimes their daughters) returned to cruising time and again: to Hawaii and back in 1934 and 1948; to England in 1935 on the *Normandie*; various ports throughout South America in 1941; to England again in 1946 and 1949, this time on the *Queen Elizabeth*; and several trips through the Caribbean in the 1950s. If the Disney Cruise Line Castaway Club existed then, Walt would be a platinum member, no doubt about it.

However, it wasn't until 1985, nearly two decades after Walt died, that The Walt Disney Company began exploring the idea of taking its beloved characters and stories to the high seas; or at least its characters.

That's when the company dipped its toe in the waters of the Bahamas, joining with Premier Cruises to offer "land and sea" vacation packages that included a few days at the Walt Disney World Resort and a few nights on Premier's "Big Red Boat" (there were actually three of them). The cruises featured special Disney-themed onboard activities for families, and, more importantly, appearances by such Disney characters as Mickey Mouse, Donald Duck, Goofy, and Minnie Mouse (who actually christened one of the ships).

The partnership lasted until 1992 and, by all accounts, was very successful, so much to the point that by that time Disney was already looking into the possibility of starting its own cruise line. And within a couple years, Disney was well on its way to designing its first ship.

Michael Eisner, then chairman and chief operating officer of The Walt Disney Company, had pushed the team to "out-tradition tradition" in the look of the ship. The result was a design that echoed the rich industry's style and past while simultaneously firmly focused on the future. What emerged was the *Disney Magic*, which set sail in 1998 and became an instant classic, a modern interpretation of such great ocean liners as the *Queen Mary* and the *Mauretania*. The ship featured a sleek, streamlined profile, with a powerful, thrusting bow, a rounded stern, and the return of something that hadn't been seen on ships in years: funnels. And, in a bit of pure Disney whimsy, the front of the ship included fanciful filigree (built around Captain Mickey), while the back featured a figure of Goofy painting the ship's name on it.

And then there were the colors of the *Disney Magic*.

"The ship was to be the colors of Mickey," according to Monica Ireland, senior project design manager, Walt Disney Imagineering, who at that time was an architect fresh out of Georgetown University. "The red and the white and the yellow were never a problem. It was always the dark color. I remember the model was painted and repainted . . . and the blue was either too black or too purple or too blue.

The Disney family was frequent cruisers, boarding such luxury ships as the *S.S. Rex* (top left) and the *Queen Mary* (bottom left). Diane, Lilly, Walt and Sharon enjoy a New Year's cruise aboard the *S.S. Constitution* (center left).

It seemed inevitable that The Walt Disney Company would eventually enter the cruise business. The first step was a partnership with Premier Cruise Lines in the early 1980s and 1990s, offering sea and shore vacations (far right).

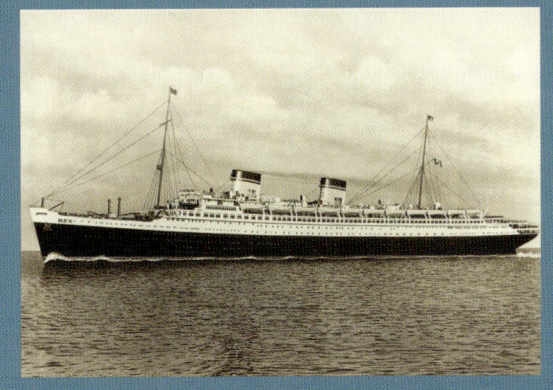

INTRODUCING THE NEW STAR/SHIP MAJESTIC

The Cruise & Disney Week.

©1989 The Walt Disney Company

The Bahamas cruise that comes with a Walt Disney World® vacation free. 7 days from $595.*

Welcome aboard for the most exciting new family vacation going — Premier's Cruise and Walt Disney World Week.

Cruise 3 or 4 nights to the Bahamas. You'll sail from Port Canaveral, just minutes from the Vacation Kingdom. Aboard the spectacular Star/Ship Oceanic, Star/Ship Atlantic, or starting in May, the new Star/Ship Majestic, Florida's biggest and best Bahamas cruise fleet. With the only true gourmet cuisine, masterpiece midnight buffets and the best entertainment going. What's more, we have extra-special stars aboard every sailing – like Mickey Mouse, Goofy and Donald Duck.

On board you'll find elegant lounges, full casinos, swimming pools, movie theatres — even full fitness programs and full-time Youth Counselors for the kids.

You have your choice of Premier's 3 or 4-night cruise and two Bahama cruise itineraries: You can either visit charming Nassau with its duty free shopping and nightlife and then sail on to beautiful Salt Cay, an Out Island paradise of glistening beaches, coral cliffs and swaying palms. Or you can sail on Premier's newest ship, the magnificent Star/Ship Majestic, to 4 spectacular Bahama Out Islands on our Abacodabra cruise to the undiscovered Abacos.

For the rest of the week, your Walt Disney World vacation is free — even on-site resorts.

Now for the magic — before or after your cruise, it's all free: If you choose the 3-night cruise, you'll have 4 nights free in Orlando. If you sail on the 4-night, it'll be three. Either way, you'll stay at one of Orlando's best hotels. (Even Disney's famous on-site and official resorts* are free if you make reservations 6 months in advance; otherwise they cost just a few dollars more.) You'll have a free Budget or Alamo rental car for 7 days with unlimited mileage, plus your 3-day Worldpassport to all the attractions at the Magic Kingdom and Epcot Center. You'll receive a free tour of Spaceport USA℠ at nearby Kennedy Space Center, too.

Either way you plan your week, rates for Premier's Cruise & Disney Week start at only $595.*

Reserve your week now. Premier's Cruise and Walt Disney World Week is available every week year-round. So send in the coupon or call your travel agent now. And ask about our round-trip Fly/Cruise rates from over 100 cities.

It all works like magic.

© 1989 The Walt Disney Company

PREMIER CRUISE LINES
The Official Cruise Line of Walt Disney World®

I'd like a detailed brochure about Premier's Cruise & Disney Week, "The Magic Vacation Combination."

Name: _____
Address: _____

Clip and mail to: Premier Cruise Lines, P.O. Box 515, Cape Canaveral, FL 32920.

*Rates are per person, double occupancy, based on published 1989 brochure rates for Super Value Season. Certain restrictions apply. Cannot be combined with any other promotion or program. Port charges not included. *On-site and official resorts subject to availability, depending upon season, certain restrictions apply. Ships' Registry: Panama, Liberia and Bahamas. © 1989 Premier Cruise Lines, Ltd.

NG 2/89

The Cunard White Star Liner "Queen Mary"

The Walt Disney Imagineering and Disney Cruise Line teams in front of the almost finished *Disney Magic* at Fincantieri Shipyard in Marghera and Ancona, Italy. At the far left in the first standing row is Bob Tracht, who has worked on every Disney ship.

Monica Ireland (opposite) still has those famous pants that gave us the deep blue color of the hull of every Disney Cruise Line ship.

"It was the end of a really long day," Ireland continued. "We're in the parking lot [at Walt Disney Imagineering in Glendale, California] and the model was on a big pedestal. It was right before a big meeting with Michael Eisner the next day. I was the one taking the notes. The team was going back-and-forth, debating. 'What are we going to do? We don't have time to repaint it.'

"Out of that conversation," Ireland recounted, "all of sudden it was [then-Disney vice president of product development] Mike Reininger who spoke up and said, 'The color of the hull of the Disney Cruise Line ships should be the color of Monica's pants.' I thought he was kidding. But everyone agreed immediately."

Her colleagues told her to bring her pants the next day to the meeting with Eisner, but she wasn't sure if they were serious or not. However, just in case, she grabbed them and stuffed them in her day bag.

"So now we're up in Eisner's office," said Ireland, "and everyone is, like, 'You brought the pants, right?' When it came time to talk about the ship's colors, we're out on the balcony of the boardroom with the ship model and they're holding up my pants to the model and saying, 'Yeah, that's it, that's the color of the hull.' I never knew until just a few years ago that the color had become known as Monica Blue."

Ireland has heard several variations of the story over the years—that it was a secretary in a dark blue dress delivering coffee to a meeting of executives, or that it was simply a designer selecting just the right color. But Ireland knows the real story and she has the proof: "I still have the pants. They're hanging in my closet. My husband says I should get them framed."

Monica Blue—and the sleek, distinctive profile, the graceful funnels, the whimsical stern characters, and countless other details—have become ubiquitous to the Disney cruise ships. The *Disney Wonder* quickly followed the *Disney Magic*, commencing operations in 1999. Both had been built at the Fincantieri shipyard in Italy.

Then, after taking a break that lasted more than a decade, the *Disney Dream* and the *Disney Fantasy*, constructed at Meyer Werft in Germany (on a scale somewhat larger than their sister ships built in the 1990s), set sail in 2011 and 2012, respectively.

Given the popularity and success of all the Disney ships, it was only a matter of time before there would be more. Sure, it would take yet another decade, but this time there wouldn't be just two ships in the class. There would be three!

And that journey would begin with the *Disney Wish*.

Monica Ireland

Project Design Manager Senior, Walt Disney Imagineering

After four years of architecture school at the University of Notre Dame, Monica Ireland had started to get a little concerned.

"I realized I did not want to do the typical architecture route," she said. "I did not want to be sitting at a CAD [computer-aided design] station drawing toilet partition details all day, so I started doing a lot of research. I wanted to find out what I could do with an architecture degree that wasn't working at an architecture and engineering firm."

What that led to was Georgetown University, where she ended up studying for her master's degree in business administration while working at the university's in-house architecture department.

"The day of graduation I happened to go to a job fair at a hotel in Washington, D.C.," she recalled, "and Disney happened to be there. The day I was walking out for my graduation with an MBA, I got a job offer from the Disney Development Company [now part of Walt Disney Imagineering] in Orlando. It was perfect timing." Ireland soon found herself on a project that was something new for Disney: a cruise ship.

"I was one of the first ones on the team," she said. "Disney Cruise Line the company didn't even exist at this point. The company was being created at the same time the ships were. This was when we were just coming up with this completely new concept of the two funnels and the colors and the references to grand ocean liners and all that."

It was also when Disney was wrestling with the question of what exactly should be on its cruise ship. "I remember when the decision was made not to do casinos and gambling," she said. "We had an entire area already designed for the casino. It was all themed around a Cuban cigar plantation as if you were in Havana. It was really, really cool, but we scrapped it all and started over. That's when the adult area came into being."

Ireland and her husband, Ralph, who was also working at Walt Disney Imagineering, left Disney in 2003 when Ralph took a job in Columbus, Ohio (near where he was from). But in 2014 they were back in Central Florida, and it wasn't long before she was consulting with Imagineering again, this time on renovations for the existing Disney ships.

Three years later, in 2017, she officially rejoined Disney as a senior project design manager on the *Disney Wish* and the other two new ships slated to follow. "I see myself between creative and project management," said Ireland. "I do more of the design processes. I work very closely with Laura Cabo, managing what the milestones are and working with the consultants."

More than anything, she brought experience to the project. "Some of my younger colleagues make fun of me because I'm 'old school,'" she said. "Reviewing drawings digitally, it's all great, but there's something to be said for sitting around a conference table with the drawings and marking them up in person."

Sharon Siskie

Senior Vice President & General Manager, Disney Cruise Line

Sharon Siskie was part of a Disney cruise line (of sorts) before there was THE Disney Cruise Line.

"I worked for Premier Cruise Lines back when they were the Official Cruise Line of Walt Disney World," she said. "So funny enough, I started in the cruise industry, then I joined Disney, and now I'm so fortunate to be able to lead the Disney Cruise Line. It really is a dream come true."

It was a dream that started in a small town in Ohio, a town so small it had one main street and one movie theater.

"I remember how excited I was to watch the Disney animated movies, like *The Aristocats* and *Robin Hood*, and the live-action ones with Herbie the Love Bug and Kurt Russell," she said. "Even then, I remember seeing the Disney name at the end of the movie and knowing it stood for something special."

She finally made it to the Walt Disney World Resort when she was 21.

"I went to the Magic Kingdom," she said, "and I remember buying a Mickey Mouse sweatshirt and feeling just like a kid again back in Ohio. I had the time of my life."

She's continued having the time of her life in Florida, earning a bachelor's degree in marketing from the University of Central Florida in Orlando, a master's in business administration from Stetson University in Deland, and over the years taking on a variety of roles in sales and marketing across Walt Disney World and Disney Cruise Line.

Today, she is responsible for the operation and overall leadership of Disney Cruise Line, including its now five (soon-to-be-seven)-ship fleet and private island in The Bahamas, Castaway Cay (with a second island destination in The Bahamas also in the works).

But she's never forgotten her roots.

"My first job in the industry was as a reservation agent," she said, "and I've always enjoyed working with guests and travel partners to provide the best vacation experiences possible for families."

Sharon is married and lives in Orlando with her husband and three cats.

"I I think my love of cats actually originated with *The Aristocats*," she said. "I still wear *Aristocat* pajamas."

Her parents also now live in Florida and, upon moving there, took on what Sharon calls "retirement" jobs . . . at Disney, of course. (They've since retired, again.)

DISNEY CRUISE LINE

The *Disney Wish* is the first of three ships joining the Disney Cruise Line fleet, bringing the total of ships to seven over the next several years.

If It Ain't Broke, Fix It Anyway

When Disney and Meyer Werft struck a deal to partner on three new ships in 2016, the strategy for planning, designing, and constructing this new class was simple: build three more versions of the *Disney Fantasy* and, yeah, go ahead and make them slightly larger. Those ships, subsequently, are a bit bigger, about fourteen thousand gross tons more than the *Dream* and the *Fantasy*, and approximately four feet longer and seven feet wider, for a grand total of 1,254 staterooms, which comes to *four* more than on those two ships (although the capacity remains at four thousand guests).

But this is Disney we're talking about, a company whose founder once said of sequels (in this case movie shorts), "You can't top pigs with pigs." This was after the overwhelming success of the Silly Symphonies animated short *Three Little Pigs* in 1933 and the pressure Walt Disney found himself under to churn out more cartoons about the pigs and their nemesis, the Big Bad Wolf. He did so (reluctantly), but the three porcine sequels only confirmed what he himself already knew (as stated above by the man himself). Not a single one of the sequels came close to matching the original, in either execution or box office figures accrued.

Imagineers learned only too well from the master, who established WED Enterprises (now Walt Disney Imagineering) in 1952 to plan, design, and build Disneyland in Anaheim, California. Sure, there are now six "Magic Kingdom"-style parks around the world, all inspired by Disneyland Park. However, because "You can't top castles with castles," not a single one is a copy of the original—or any of the others for that matter.

In fact, Imagineers like to tell a joke about themselves and their way of working that first appeared in the book *Walt Disney Imagineering: A Behind the Dreams Look at Making the Magic Real*:

Question: *"How many Imagineers does it take to change a lightbulb?"*

Answer: *"Does it have to be a lightbulb?"*

Of course, the *Disney Wish* was always going to be a cruise ship, and certainly no one was going to mess with the distinctive profile and defining colors that have made the Disney Cruise Line fleet the most recognizable on the seven seas (and all the oceans, too).

But what about the rest of the ship? Could it be something different?

"It would have been very easy to just build another *Disney Fantasy*–level ship," said Principal Show Manager Bob Girardi, an Imagineer who previously worked on that ship and the *Disney Dream*. "But what's the challenge in that? We looked at it and said, 'We can do a few things better.'"

"It's natural," said Steven Read, marine, technical, and engineering planning executive, Walt Disney Imagineering, and a veteran who has been involved in all the Disney Cruise Line ship builds. "[With] anything in life, you want to improve on your previous version, whether it's as simple as writing an email or a document, or as complex as designing and building a cruise ship."

Former Imagineering President, now Global Imagineering Ambassador, Bob Weis, was willing to be even more radical than that.

"This was our opportunity to do something truly unique," he said. "You don't very often get a chance to go back and rethink all your basic precepts. Everyone who's been involved in these next ships has been, 'Hey, let's throw it all up in the air and do something new. Let's be willing to learn from our experience.'"

To that end, a team of executives and designers from Disney Cruise Line and Walt Disney Imagineering spent several months in fact-finding mode in 2016, evaluating the latest developments in the cruise line industry, tracking trends, and what advancements were being made, plus weighing feedback from guests and gauging what their expectations might be for the new ships. We even embarked on a top-to-bottom assessment of what was working—and what wasn't—on the existing vessels.

Meghan Moore, experience integration project manager, Disney Cruise Line, pointed to one item in particular that the team wanted to take a look at. "We knew we had no problem filling the restaurants during meals and the theaters during shows," she said, "but how do we maximize the use of the spaces in between those times?

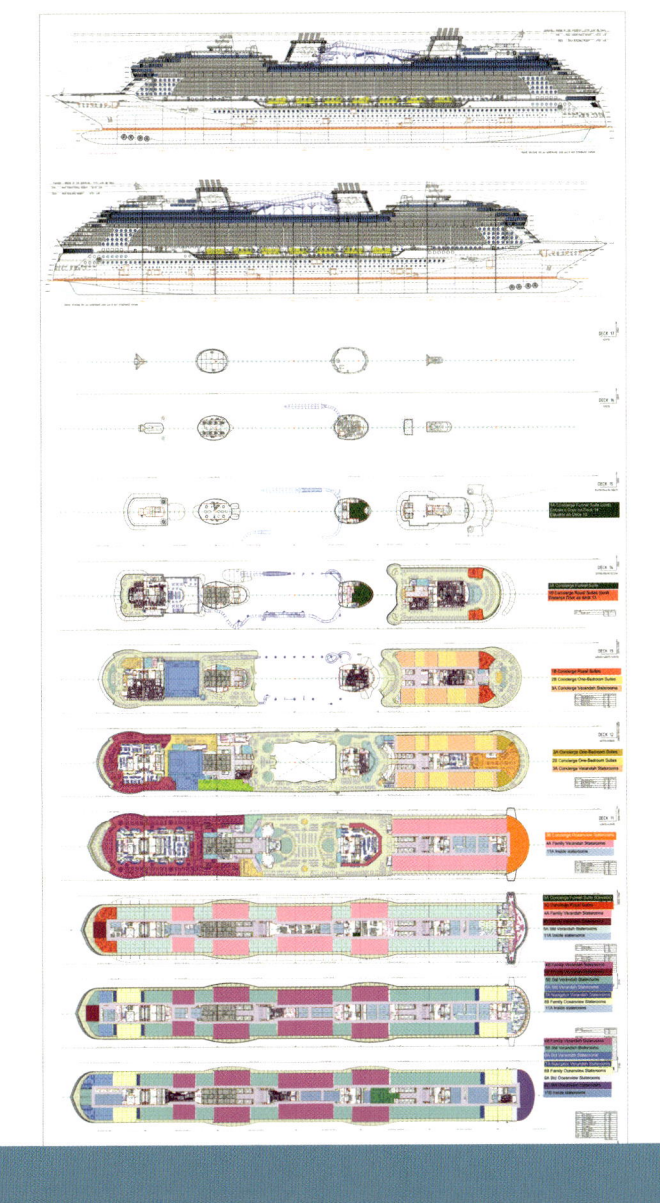

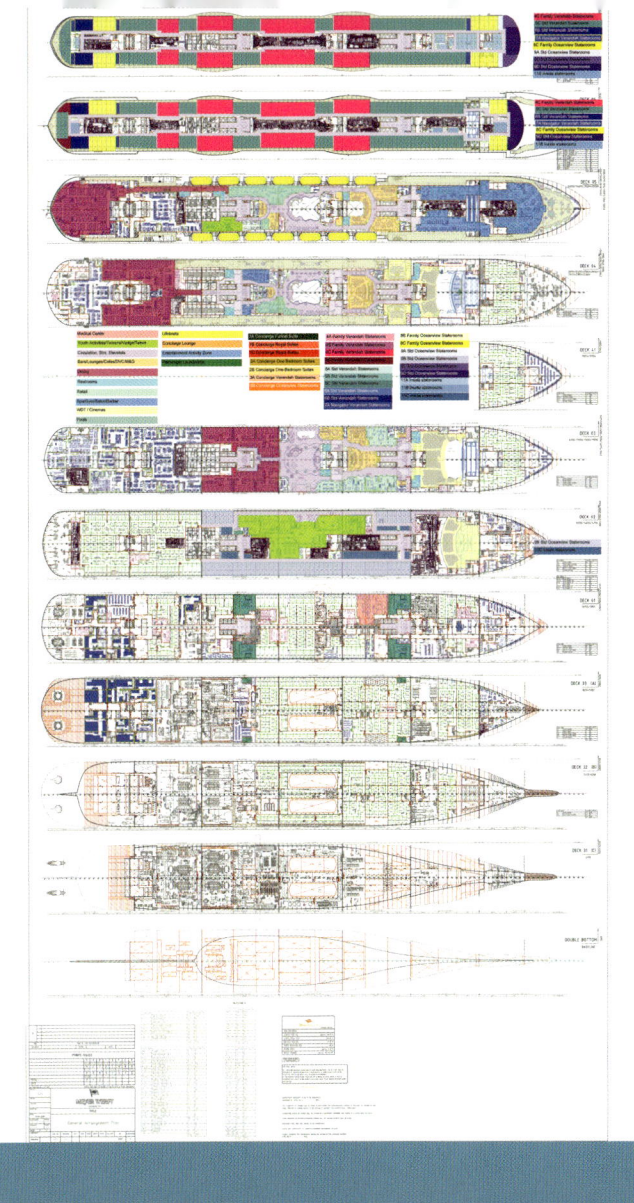

The GA, or General Arrangement, was used to design the layout of the *Disney Wish*. It's essentially a finely detailed map, indicating the location of every conceivable space on the ship.

Wall Pattern

Elevator Doors

Floor Design

(Opposite) An early model of the *Disney Wish*. (Not really. It's actually a cake baked by Anna, Emma, and Vera, the daughters of Portfolio Project Management Executive Philip Gennotte.) A computer model (below) of the new Deck 2 location of Disney's Oceaneer Club, showing the slide from Deck 3. The forward and aft elevators and stair landings (left) have been expanded to accommodate more guests.

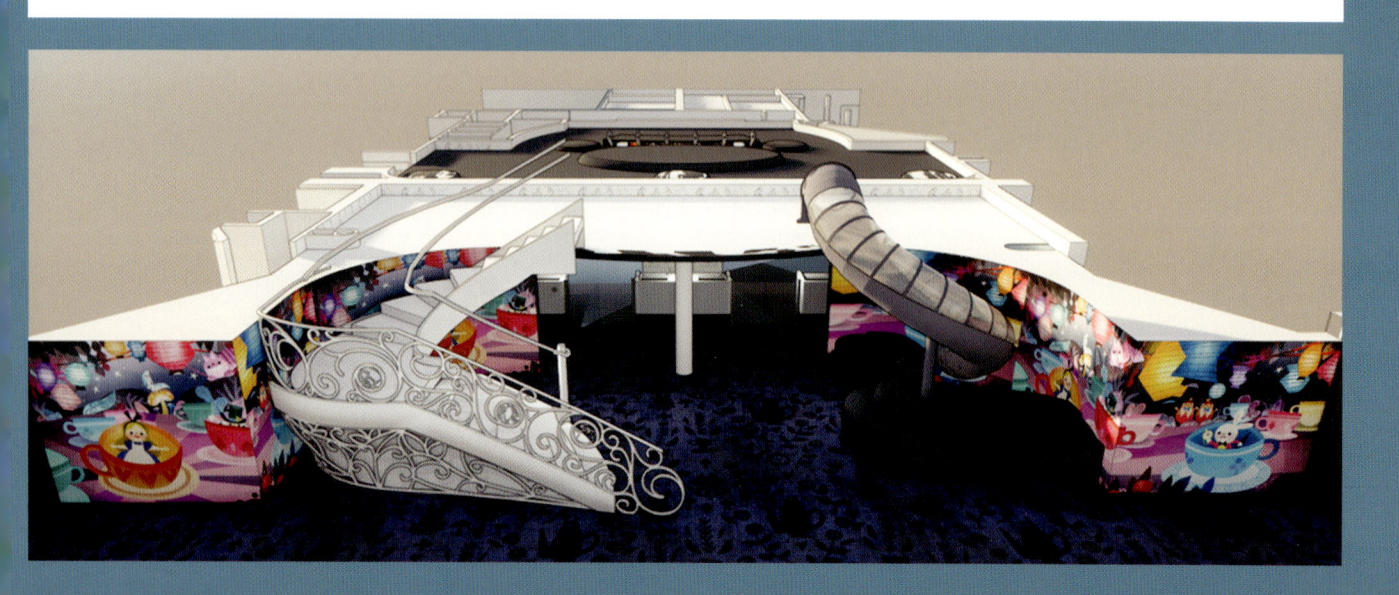

It's a bit like putting a puzzle together, only you don't know what the puzzle looks like yet because you're sort of making it up as you go.

"We had an understanding where the theater was and where the restaurants would be, and of course the accommodation spaces, but no real understanding of where the lounges, the stores, the spa, or the upper decks would be; nothing was arranged," said James Willoughby, director, hotel operations and special projects, Disney Cruise Line. "The first six months was that period of 'boxology' and really tying down the idea of where, geographically, everything would be positioned for efficiency and guest experience."

"Boxology?"

"'Boxology' is what are the different functions, what are the different spaces, and how are we going to fit them in the way that our partners at Disney Cruise Line want on the ship," noted Atwood. "Once we established that, we were able to start with the structural and mechanical plans."

"Also," she added, "we were concerned that some areas of the ship currently seem a bit out of the way, so the question becomes [this]: How do we better integrate them into the flow of the ship so they don't seem so hidden?"

Enter the General Arrangement plan.

"Just as a resort has a building footprint, so does a cruise ship,' said David Atwood, principal project manager, Walt Disney Imagineering.

"It's called the GA. Instead of looking at the plan floor-by-floor as you would with, say, a resort hotel, we're looking at it deck-by-deck."

The GA is a drawing that serves as a layout for the entire ship, defining everything from the spaces, compartments, and bulkheads to the shape of the hull, the volume and number of decks, and the location of all the equipment, regardless of function.

The GA and its "boxology" led to some dramatic changes in the way Disney cruise ships are normally laid out. These include:

- The redistribution of bars, lounges, and a new-to-the-Disney Cruise Line entertainment venue to increase their visibility and accessibility.

- The relocation of "it's a small world" nursery and Disney's Oceaneer Club to Deck 2, conveniently situating it below the ship's main hall.

- The repositioning, separation, and expansion of the spa and fitness facilities to Deck 5 (along with a new outdoor space), which opened the area they previously occupied to the first forward-view staterooms on a Disney cruise ship.

- A reimagined sports court that is now indoors, climate-controlled, and able to host a wider variety of events and activities.

- More pools, more water amusement options, and better views on and from the upper decks.

- And, more subtly, reducing the number of elevator banks and stair halls from three on the existing DCL ships to two on the *Disney Wish*. (And don't worry: there are actually more elevator bays in total on the *Disney Wish* than on any other Disney liner. They're just distributed a bit differently to provide better flow throughout the ship.)

The removal of the midship elevators and stair halls was a boon to those contending with the ship's main hall design, who were given an opportunity to make the area much larger—and include a stage, to boot.

"Going from three elevator shafts to two elevator shafts changed everything in terms of our design of the Walt Disney Theatre," said Bob Girardi. "We were designing it as a 1,350-seat theater, which it is on the *Dream* and the *Fantasy*, and now we have to work around a larger bank of elevators because the midship elevators have been removed from the plan. And the forward elevators that border the theater just got bigger. We finally got the theater to where it is today at about 1,275 seats, give or take."

So what began as an exercise in designing and building "the *Disney Fantasy*, just ten percent bigger," according to David Atwood, had now resulted in something that was vastly different than what Guests were experiencing on the existing cruise ships.

"We had played with that GA for over a year and a half, trying to get it just right," said Pam Rawlins, executive producer, Walt Disney Imagineering.

It was time to move the ship from theoretical concept to actual reality.

And that's when Philip Gennotte and Laura Cabo came aboard.

24 Making a *Wish*

Laura Cabo and Philip Gennotte in the garden of Philip's home in Leer, Germany, shortly after taking over the project (right).

Laura, Anne Marie Gullikstad of YSA Design, and Claire Weiss discuss themes and designs for spaces on the ship (far right top).

Philip chats with Lauren Fong in the construction hall at Meyer Werft (far right bottom).

A cross-section of the ship, providing a sampling of what's in the various spaces. A more detailed color in version of this art appears in the Grand Hall.

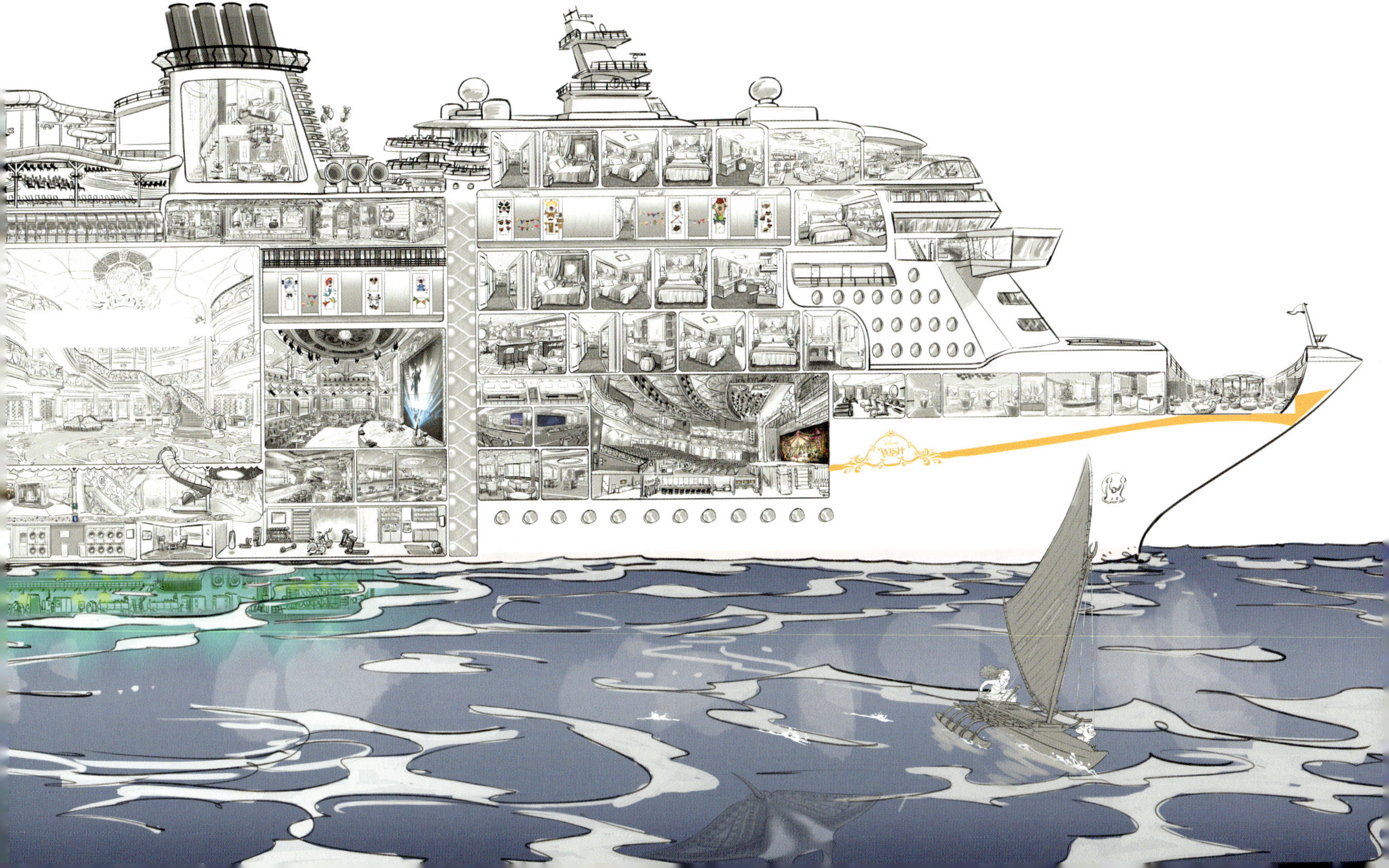

Meghan Moore

Experience Integration Project Manager, Disney Cruise Line

On a visit to Magic Kingdom Park when she was a little girl, Meghan Moore was walking down Main Street, U.S.A., when she turned to her grandmother and blurted out, "I'm going to work here someday."

Meghan wasn't sure what she was going to do when she eventually worked at the Walt Disney World Resort, but she was certain of one thing: it wasn't going to be as an engineer.

"My dad and my brothers are engineers," she said, "so I actually refused to be an engineer. I had to be different."

She entered Iowa State University as a computer science major because she loved math and science. But that's not the discipline in which she earned her degree. It was industrial engineering.

"Industrial engineering is really the perfect balance of business and engineering," she said. "It has endless possibilities on where and what you can do with the degree."

What about that vow to not become an engineer? Not only did she get over it, she embraced it.

"Today, I'm proud to say I'm an engineer," she said, "just like my dad and my brother."

She's also proud to say she finally ended up working at Disney, first at the Walt Disney World Resort in Florida and now as part of the team in Papenburg, Germany, where the *Disney Wish* was built.

One vow Meghan was more than happy to make was with Albert Mollema.

"We met the traditional way—with an online app," she joked.

Albert lived in Groningen, the Netherlands, about an hour's drive from Papenburg. For their first date, which took place in early 2020 right before the pandemic shut everything down, the two spent eight hours walking around his hometown.

The relationship took off from there and Albert moved to Germany in 2021, landing a job at the warehouse that supports the construction of the three Disney ships.

Today, they have their own Disney wish, one in which they live happily ever after.

David Atwood

Project Manager Principal, Walt Disney Imagineering

David Atwood was a semester away from graduating with a degree in neuroscience from the University of Florida. Medical school was in his future.

"I was studying one night in the library at Shands Hospital and just couldn't get focused on studying for my final," he said. "Literally, like a ton of bricks hitting me on the head, it was like, 'What am I doing? Why am I doing this?' Growing up, I'd always done really well in school and aced all the tests I'd taken and been convinced that, 'Hey, you can do this.' I thought becoming a doctor would be a great career path and it is if you're passionate about it, but I'd never been passionate about it."

As he sat there, pondering his future, he happened to glance at a rack of magazines and noticed a copy of *Architectural Digest*.

"I'd always loved drawing cars, buildings, airplanes, spaceships, and things like that when I was growing up," he said. "I actually won a few art competitions, including one in which my painting was reproduced on a billboard in Jacksonville Beach, Florida, with my name on it."

But David decided to pursue medical school because it seemed like the thing to do—and he knew he was up to the challenge.

Until he spotted that copy of *Architectural Digest*.

"I just picked it up and started flipping through it. That's when I realized that this was what I really wanted to do all along. I pulled the trigger and decided to change my major."

Three years later, he finally had the degree he wanted—in Construction Management.

After graduation, he went to work for a contractor that was involved in the construction of Disney's Animal Kingdom and was hooked—on Disney.

"I was blown away by the level of detailing and theming," David said. "I knew one day I was going to be an Imagineer.

That day didn't come for more than a decade. However, since joining Imagineering in 2007 as a project manager, he's overseen the construction of Disney resort hotels in Florida and Shanghai—and now Disney cruise ships in Germany.

"Building a cruise ship is so unique" he said. "It's easily one of the coolest things I've ever worked on."

The theme of the *Disney Wish* is built around enchantment, inspired by classic Disney stories and characters, among them Cinderella.

The Theme of It All

It was the week of February 5, 2018, that the design kickoff for Project Triton at the convention center at Disney's Yacht & Beach Club resorts at the Walt Disney World Resort in Florida was launched. All the principals were there: the Imagineers who had developed the initial concepts for the ships, plus their partners at Disney Cruise Line who had provided input on programming and operations, and the external design firms involved, including Jeffrey Beers International, out of New York; Tillberg Design of Sweden; YSA Design in Oslo, Norway; and SMC Design from London.

Specifically at the meeting as well was Laura Cabo, who at that time was heading up the architecture studio at Walt Disney Imagineering.

"I wasn't even on the project," said Cabo. "I was there as a favor to [Walt Disney Imagineering president] Bob Weis to just look in and see what was happening."

Across the room was another project newcomer, Philip Gennotte, who had worked on the *Disney Dream* and the *Disney Fantasy* but as a project manager for Meyer Werft, the shipbuilding company that had constructed the previous two vessels. He had just been hired by Disney a few months before and wasn't expecting to be working on the *Disney Wish*.

"It wasn't the plan to bring me on to the project," he said. "I wanted to stay away from the new ships because I didn't want any conflicts of interest [from my previous association with Meyer Werft]. I joined Disney Cruise Line as a true cruise ship build-and-design expert who was going to run the site office."

But plans change, especially after Bob Weis boarded a flight for Orlando in early 2018.

"I got on the plane, went to my seat, and sitting next to me was Wing Chao, a Disney Legend who was deeply involved in the first generation of Disney cruise ships," said Weis. "At some point on the flight, I said, 'Hey, Wing, let's talk business for a second.'"

Weis told Chao how he was looking for someone to bring an entirely new perspective to the way Disney designs its cruise ships.

"I asked him, 'Where can I go to find this person?' And he said, 'You already have this person.' I said, 'I already have this person? I have a lot of talented folks and I haven't figured out who it is. Who is it?' He said, 'You have her. Her name is Laura Cabo.' I said, 'Laura? She's very important to us, but she's running the architecture department.' Wing said, 'I think you should talk to her. I think she could be the person you're looking for.'"

Turns out, Wing Chao was right (which is why he's a Disney Legend).

"Within three weeks, Laura and the team she assembled had completely reinvented what our view of a cruise ship was," said Weis. "I think they did an incredible job of rethinking Disney and the cruise visitor and how guests are going to experience these new ships."

By summer Cabo and Gennotte found themselves leading the Walt Disney Imagineering *Disney Wish* team, with Cabo as the portfolio creative executive in charge of the design team and Gennotte as the portfolio project management executive responsible for delivering the ship.

"We both came from outside the project," said Cabo, "which I think is a benefit because it gave us a fresh way of looking at things. We didn't fall into the pattern of, 'Oh, this is how it was done before.' Philip had extensive cruise line experience, but he was relatively new to Disney, and, though I was eight years in, this was the first time I was really leading a creative project for Imagineering."

It was also the first time Cabo had ever worked on a cruise ship, which meant she had to get up to speed—and fast. The best way to do that was to sample the product: she booked passage on a couple of Disney cruises. That was yet another new experience for her, and it showed.

"My first impression of Laura occurred during a safety drill, where everyone has to go to their assembly stations," said Imagineering creative director Danny Handke. "She was in the elevator lobby, and saying, 'I don't know where to go.' Sachi [Danny's wife, who is also a member of the *Disney Wish* project team] and I helped her [figure out] where she needed to go, but I think that really stuck with her because the wayfinding became so important to her, making sure that everything was just clearer and better defined."

While Cabo was getting familiar with the Disney ships—and already making notes about what she hoped to

32 Making a *Wish*

Walt Disney Imagineers drew on Disney's rich history of "fairy tales and castles" in developing ideas and artwork for the *Disney Wish* (bottom right). They turned to one story in particular, *Tangled*, for the character that would join the lineup of iconic Disney characters that appear on the sterns of the existing ships. Rapunzel, seen here in an early sketch (top) and as a scale model (bottom left), is the first Disney princess to get the nod.

Philip Gennotte

Portfolio Project Management Executive, Walt Disney Imagineering

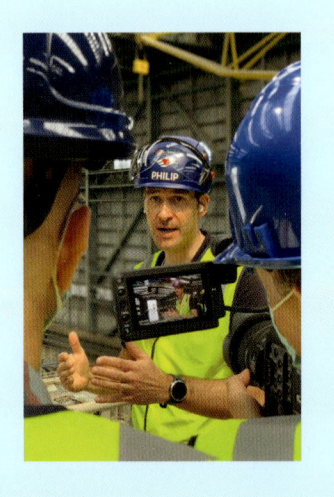

Philip Gennotte has always been a model builder. From a very young age, he was fashioning planes, boats—especially ships, not only from kits, but on his own with wood, cardboard, and whatever he could find.

"I would even write to shipyards and say, 'Hey, I'm interested in building this ship' and in the old days shipyards were very supportive of model builders and would just send drawings free of charge."

Philip was also fascinated with Disney. "As a child, you always grow up with it," he said. "Even in Belgium, where I grew up, Disney was something, of course. Mickey Mouse, Donald Duck, I always watched the cartoons.

"My grandparents took me to the cinema for my first time ever, in Antwerp, to see *Snow White and the Seven Dwarfs*. When I look back on it now, I think how unbelievable that was, that I saw Disney's first movie as my first movie. It left an impression for life really.

It seemed inevitable, then, that this builder of model ships and this fan of Disney would one day end up putting both of those passions together to manage the construction of Disney cruise ships.

It took a while, though. The Disney, not the shipbuilding.

After graduating from the University of Newcastle upon Tyne (UK) with a degree in Naval Architecture, Philip went straight to Meyer Werft in Papenburg, Germany.

"I started out as a true naval architect, designing ships," he said. "I actually worked with the Disney team on developing the layout of the *Disney Dream*."

But it was a trip to Walt Disney Imagineering headquarters in Glendale, California, that convinced Philip that maybe his future would be with Disney.

"After the *Disney Fantasy* had been delivered [in 2012], I had the opportunity to do a presentation on the ships to Imagineers from the shipyard point of view," he said. "Now I'm a passionate model-builder, so I just had to see the workshop where they build the models. I was taken there and just told to enjoy.

"How awesome was that, right? I got an hour to just wander around and take it all in."

Today, Philip takes it all in as the Walt Disney Imagineering portfolio project management executive for the team designing and delivering the *Disney Wish*.

"Even today, I keep wondering to myself, is this even possible? How do we do this? I'm pinching myself. This is incredible."

change—Gennotte was busy tinkering with something that everyone thought was already locked in.

"I spent the first few months fine-tuning the General Assignment plan (GA) to make sure it worked," he admitted. "I was doing a lot of sketching. One of the things we discovered is that, although we had expanded many of the spaces on board, we hadn't accounted for the fact that we were going to need more crew members to staff them.

"Because of that, we had to find a way to get more crew cabins," Gennotte pointed out, "so we could accommodate them."

Likewise, Cabo, fresh off her Disney cruising experiences, was ready to jump in as well. In the weeks after the design kickoff meeting at the resort in Florida, Cabo found herself increasingly busy with Project Triton. She sat in on a number of brainstorming sessions and began to map out her ideas for what the new ships could be and what features might be on them.

Funny thing, though, was, she still wasn't on the project.

"I wasn't officially onboard yet," she said. "I was just working on it, part of my time."

Not that she wasn't ready, willing, and able.

While Cabo was unfamiliar with cruise ships, military vessels were another story. "Growing up, my dad was a career military officer on big ships, so I spent a lot of time as a little kid wandering around ships that he was captain of," Cabo recollected. "He brought back many stories and a lot of things from around the world. That's what Walt Disney did. He brought those wonders of the world back here and created his theme parks. And then we've gone on to create these cruise ships, which open a portal to the rest of the world for our guests. I wanted to be a part of that."

She would get her chance, maybe a little sooner than she anticipated.

On April 18, 2018, just two months removed from her introduction to the project (as an "observer"), there she was in front of Disney's "Big Cheese"—and we're not talking Mickey Mouse here.

"I remember the first time I presented to Bob Iger," she said. "I was so nervous."

In fact, all the Bobs were there: then-chairman and chief executive officer Bob Iger; then-chairman of Disney Parks, Experiences, and Products (and now chief executive officer) Bob Chapek; and Imagineering president Bob Weis.

And here was Cabo, already expounding on what the vision could be for the new cruise ships.

She need not have worried. Everyone loved her take on the design—and soon enough she got her own wish: to lead the creative effort for the ships, which at that time were known by the very *catchy* names of Ship 5, Ship 6, and Ship 7.

Deck Four 35

Meanwhile, Gennotte, who was still working on the layout of the ships and crunching the numbers on how many staterooms there should be, finally got the nod as the portfolio project management executive. He saw what Cabo and her team were doing . . . and he was excited.

"To come up with a theme, that has never been done before, even on the previous Disney ships was a thrill," he said. "*Dream* and *Fantasy*, all that was said was that they were Art Deco and Art Nouveau. As beautiful as they are, that's not really a story. What Laura was doing was revolutionary in my mind and really drove the project from there onward."

By the summer of 2018, Cabo had pulled together a team at the Walt Disney Imagineering campus in Glendale, California, in a building called the Bowling Alley (because that's what it actually had been), and essentially hit the reset button.

"I felt there was a huge opportunity to connect in a bigger way to our stories and characters," she said. "We sat down in a conference room and scribbled down a bunch of ideas. We began thinking about these three ships as each having its own personality. We didn't want to pick something out of an architecture book and say, 'The style of this ship will be mid-century modern and that one will be Gothic.'

"We needed something that was more emotive," Cabo added. "Enchantment was always there, right from the beginning, so we knew that would be the first new ship.

"It sprang from the idea of a castle," she continued. "A castle is the centerpiece of our parks, so it was really kind of obvious to us that we should give Disney Cruise Line its own castle on the seas."

Also obvious, at least to Cabo, was what the name of the ship should be.

"As we focused in on enchantment as our theme," she said, "we looked at the incorporation of Cinderella, our most iconic princess. At the center of her story is 'a dream is a wish your heart makes.' There it was. We just had to put 'wish' on there."

From there, the idea swept through the ship like so much pixie dust.

According to Pam Rawlins, the executive show producer, the team had already been looking at various motifs, such as "fairy tales and castles" and "forests and animals," that would eventually end up in the final plans, "but it was Laura [Cabo] who brought those stories together under a unified theme."

"The design is so intrinsically linked to the creative," said Cabo. "By changing the GA and developing the theme, we were able to weave it throughout the whole ship and create this amazing series of environments and experiences for our guests."

Which set the stage for another significant change Cabo wanted to make.

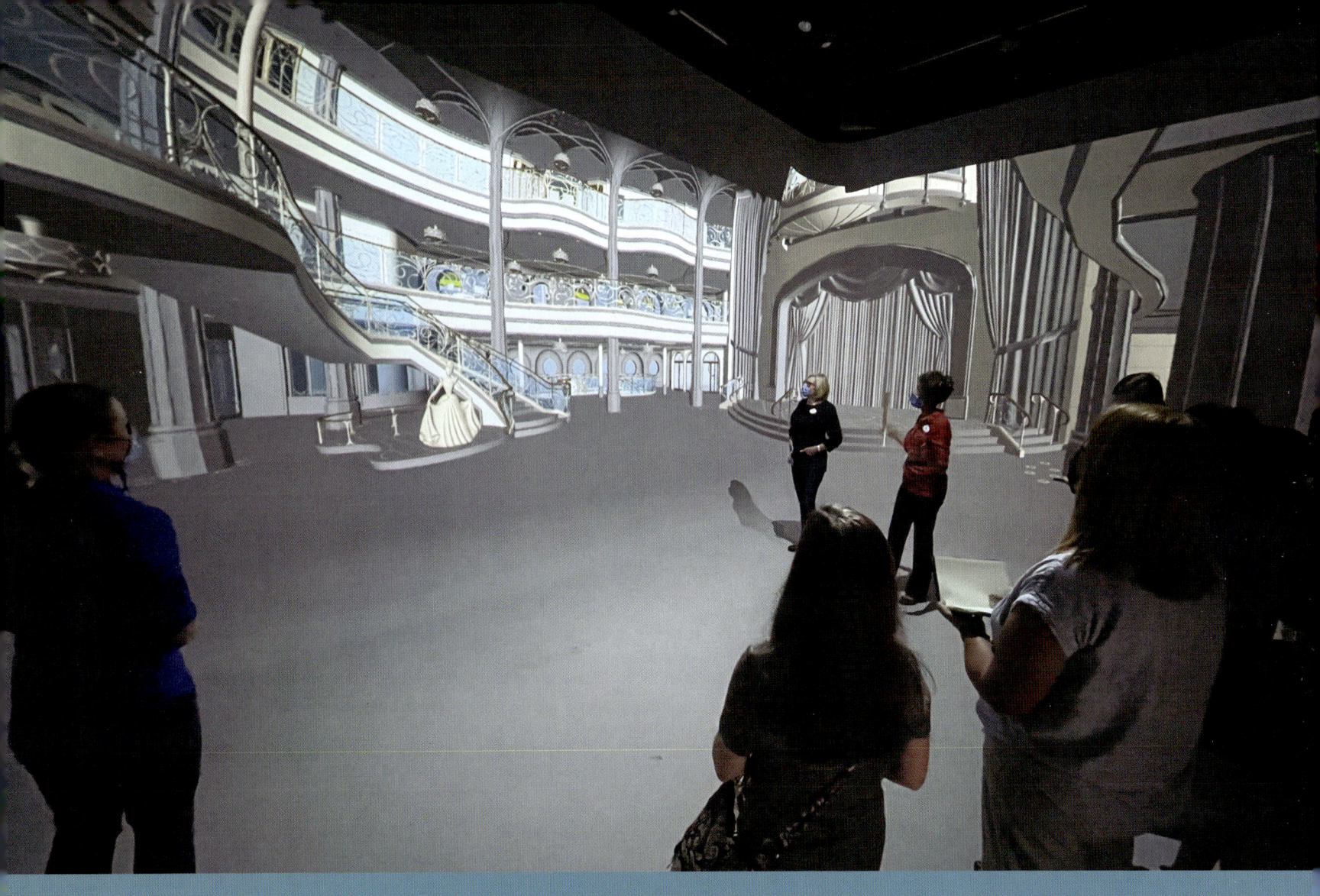

That's not the actual Grand Hall, but an incredible simulation created in Walt Disney Imagineering's Digital Immersive Showroom, also known as the DISH. The DISH is a pre-visualization tool that enables designers to virtually inhabit a space before it's built.

Laura Cabo

Portfolio Creative Executive, Walt Disney Imagineering

Unlike many of her colleagues, Laura Cabo did not grow up wanting to work at Disney, but that doesn't mean she didn't like Disney.

"We were typical Americans," she said. "As a kid, I remember on Sunday nights watching *The Wonderful World of Color* on our old black-and-white TV. My brother and I would be in our pajamas because we would have to go to bed right after it was over."

She also recalls going to Disneyland while her U. S. Navy captain dad was stationed in San Diego.

"People in the military and their families got free tickets to Disneyland," she said, "so we would drive up to Anaheim every year we were in California and it was the highlight of the year."

But when it came time to embark on her career in architecture, Laura chose a small design company in Boston that—wait for it—ended up doing a lot of work with Disney.

"In 1991, our little firm won a competition to design Disney's first large-scale moderate-priced hotel at the Walt Disney World Resort, Coronado Springs," she said. "For our presentation, we created a story and dressed up in colors that reflected the theme of Mexico and we had all these festive, brightly colored lights in the conference room. We pulled out all the stops and won the contract."

That Disney job led to other Disney jobs, including developments in Marne-la-Vallé, France, and Celebration, Florida.

Eventually, inevitably, Disney itself came calling.

"I was skeptical about coming to Imagineering at first," Laura said. There was so much I didn't know about, which is all of the Imagineers working on the attractions. I always worked on the resorts. I didn't know about what else was going on 'behind the curtain.'"

But her biggest challenge in accepting the position turned out to be something—or, more accurately, someone—closer to home.

"The hardest part was talking my Bostonian-born-and-bred, Celtic-loving husband into Laker-country," she admitted.

When Laura arrived at Imagineering in 2013, she took over management of the architecture and interiors studios, but design was never far away.

"If you told me I was going to be working on the cruise ships, I never, ever would have guessed that, ever," she said. "But you just fall in love with your team and your project and throw yourself into it. It really is a magical thing to work with such talented people."

Who better to represent a ship built on wishes than the Disney princess who sings about them? Cinderella, seen here in a foam sculpt, may be the most prominent figure in the Grand Hall, but she certainly isn't the only one. Lucifer and a couple of her mouse friends, Gus and Jaq, also make appearances.

Cinderella

Wish fulfillment is at the heart of many a Disney story, whether it's wishing upon a star, wishing for the one you love, or wishing with all your heart.

The latest Disney ship is all about the dreams and desires of Disney characters—and the bit of enchantment—that can turn their once-upon-a-times into happily-ever-afters.

Who better, then, to represent the spirit and theme of the *Disney Wish* than perhaps the most iconic Disney princess when it comes to dreaming and wishing and having heart (she sings a song about them, in case you haven't heard)?

We all know the story by now. Cinderella's mother died when she was young. Her father remarried, exchanging vows with Lady Tremaine, who brought her own two daughters into the family. When Cinderella's father died, Lady Tremaine took over the house and the household, and Cinderella was relegated to the role of scullery maid and serving girl.

To her credit, Cinderella made the most of her hard-knock life (oops, wrong orphan story). Despite the increasingly harsh treatment she received at the hands of her stepfamily, she remained as she always had been: a kind, loving girl, who not only fed the mice that lived in the house, but sewed them little outfits, too.

What sustained her were her dreams and her wish (cue that song again) that one day she would find true love, be whisked away from her evil stepfamily, and live happily ever after.

Spoiler alert: her wish comes true when the Fairy Godmother appears and whips up a beautiful dress and turns all the animals around the house into horses and valets so she can go to the ball, dazzle the prince, bibbidi-bobbidi-boo.

This story of cheerful perseverance in the face of cruel and severe circumstances makes Cinderella the perfect embodiment of the *Disney Wish* (not that that is the story of the *Disney Wish*, mind you). Fittingly, she appears in the Grand Hall in her beautiful blue ball gown (well, imagine it's blue—the statue of her is actually a bronze), as sweet and happy and joyous as the day she was born (and, to her at least, every day since then).

In fact, the fairy tale–inspired Grand Hall is a tribute to the entire story of Cinderella, including not just the bronze statue of the beloved princess, but also the stained glass embellishments in her signature colors, and icons from the film in the carpeting, metalwork, and light fixtures.

It truly is a place where she can—with delighted guests all around her—live happily ever after.

Rapunzel

Rapunzel is not an Imagineer, but her artistic talents certainly could make her an honorary one.

After being kidnapped as a baby and raised in a tower, she used that "alone" time to develop some pretty serious creative skills, including painting, singing, dancing, and working on many different crafts.

And now, she's left her tower—and seemingly the world of the film *Tangled*—for the stern of the *Disney Wish*, using her talents to help out with painting the filigree and the ship's name.

This is no small order—and neither is Rapunzel, or her chameleon pal Pascal, who is lending her a palette as she executes her swirls and flourishes.

Rapunzel is eight feet, 10 inches tall and her hair, stretching across the back of the ship and holding her in place, is over 59 feet in length.

But it takes more than hair to keep her there.

Strong winds, the salt of the sea, and even the vibrations of the ship can take a toll on one's health, well-being, and looks, so she's been securely fastened to the ship (in artful and clever ways that aren't immediately apparent) and coated with polymers that keep her fresh, shining, and lively—or least lively looking.

Rapunzel continues a tradition of Disney characters on Disney cruise ship sterns that started with Goofy on the back of the *Disney Magic* and also includes Donald Duck on the *Disney Wonder*, Sorcerer Mickey Mouse on the ***Disney Dream***, and Dumbo on the *Disney Fantasy*.

Which characters will be on the next two Disney ships?

We *wish* we knew. (Well, we do, but we're not telling yet.)

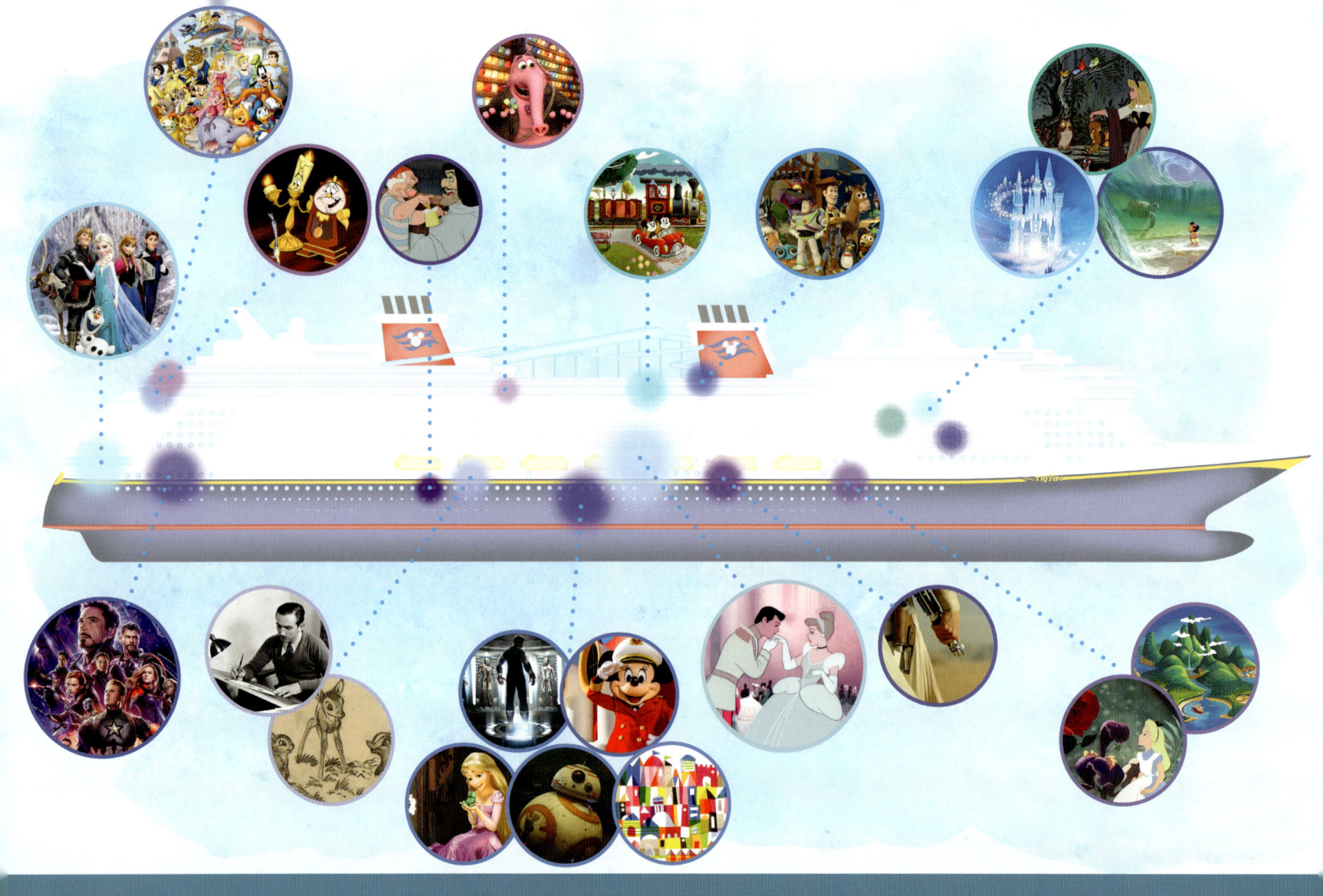

Imagineers put together a map to track the various stories, characters and even people that are appearing on the *Disney Wish*. "We wanted to make sure we had a good mix of Disney storytelling," said Laura Cabo, "from princesses and pirates to Marvel and *Star Wars*. There's something for everyone here."

Stories and Characters and Motifs, Oh My!

How much Disney should a Disney cruise ship have if a Disney cruise ship could have Disney? It's a convoluted question Walt Disney Imagineering and Disney Cruise Line have been wrestling with since the *Disney Magic* first sailed into Florida's Port Canaveral in 1998. With the first four ships, the designers had always hedged their bets. Sure, the ships were filled from bow to stern with Disney characters, shows, spaces, and references. But there were also places where you could take a break from all Disney all the time. These spaces included Cabanas, the food court that recalls a breezy boardwalk along a picturesque Australian beach; PALO, the premier dining experience with elegant, northern Italian-inspired décor; and the Cove Café, the adults-only lounge serving specialty beverages (and a welcome dose of peace and quiet).

That philosophy started to change in November 2016 when Parrot Cay, one of the three original rotational restaurants on the *Disney Wonder*, ditched its Key West-meets-Jimmy Buffett, very non-Disney interior design for a swinging, New Orleans-style supper club inspired by *The Princess and the Frog*.

Tiana's Place, with its live jazz and character appearances, was so popular with guests that soon the previously neutral Promenade Lounge just outside its doors was also given a makeover, transformed into French Quarter Lounge, a pre-show of sorts for the restaurant and its raucous atmosphere.

As Laura Cabo arrived on the scene, the Imagineers working on the new ships were having the same discussions that their predecessors had had on the existing ships.

"It was really clear that the idea of going a lot more into an environment that people loved and recognized,

Deck Five 43

loved the music, loved the characters, that those were things that really differentiated us," said Bob Weis. "That was really the key to what these next ships are going to be: let's push story a lot harder."

As Laura Cabo arrived on the scene, the Imagineers working on the new ships were thinking the same thing.

"At the beginning of the project, our thought process was a little different," said Claire Weiss, who, along with Danny Handke, served as a Walt Disney Imagineering creative director on the *Disney Wish*. "We started with this idea that guests may want a break from Disney, that they need spaces where they're not part of the Disney universe. The more we designed and thought about it, the more we started to think we might not have it right.

"People are on a Disney cruise for a reason," Weiss pointed out. "They want to feel part of the Disney experience and there are ways to do that subtly and beautifully."

Added Pam Rawlins, "Some people thought we were going to be too heavy with Disney, but Laura was instrumental in convincing everyone that Disney stories and motifs could be done at a sophisticated level. We're really setting the bar here."

One person she didn't really need to convince was an executive who seemed like a pretty good ally to have in your corner. "We were reviewing our concepts with Bob Iger [who was chairman and CEO of The Walt Disney Company at the time]," Rawlins recalled. "We had all these beautiful boards with our themes and how we were going to interpret them throughout the ship, and he just kept saying, 'More Disney, more Disney.'"

This meant not only purely "Disney" stories, characters, and settings, but everything in the Disney universe, so to speak: Pixar Animation Studios, Marvel, and *Star Wars*. He was essentially telling the design team that they shouldn't hold back when it comes to using Disney properties on the ships, whether it's in a restaurant, a café, or even a bar geared toward adults.

Cabo and her team had a clever way of handling the amount of "Disney" that would be found in different types of environments aboard the ship. "Laura was instrumental in breaking the design into two kinds of spaces: motif spaces and story spaces," said Danny Handke.

The vessel's other creative director, Weiss, explained further: "A story space is one where we have a time and place story that's usually based after a film, like *Frozen* or something in the Marvel Cinematic Universe. It's really clear. We're taking you into this world.

"The motif spaces," she continued to explain, "are celebrations of familiar environments, like, 'This is sort of like a section of a ship from Norway, or this is sort of like a castle environment.' They're tied more to an idea or a story than to a specific film.

"The Grand Hall is a good example where it's inspired by *Cinderella* and, design-wise, it is completely immersive.

Imagineers developed three new concepts for the signature restaurants aboard the *Disney Wish*. Sketches and renderings include the Arendelle grand dining hall (top), as well as the entrance to the restaurant (bottom left) and a corridor leading to the reception area (bottom center); a dining area in 1923 (middle); and the Worlds of Marvel entrance (bottom right) and reception area (left, second from top). The Worlds of Marvel team (right, second from top) meets around a scale model of the restaurant.

Concept sketches

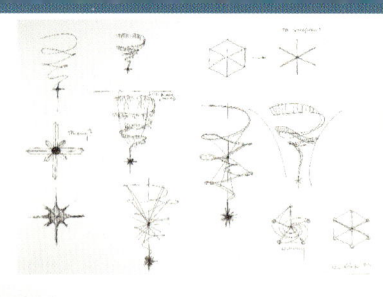

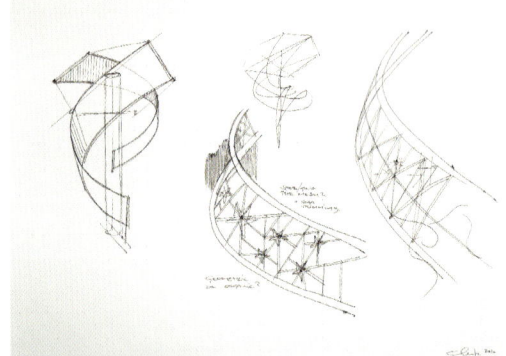

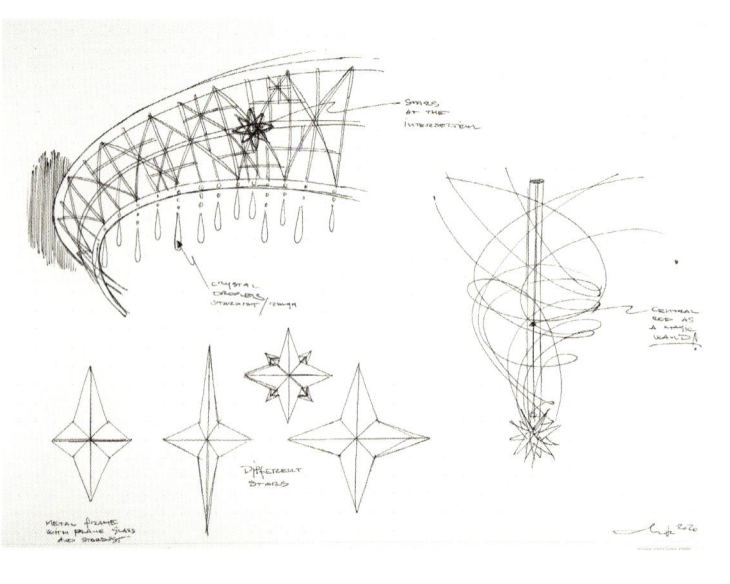

The song "Oh Sing Sweet Nightingale" from *Cinderella* not only inspired Nightingale's, but the design of the chandelier inside the lounge. From a loose sketch of soap bubbles and musical notes (top left), designers fashioned an intricate light fixture that also features a crystal nightingale at its center (bottom right). (It's in there. You just have to find it.)

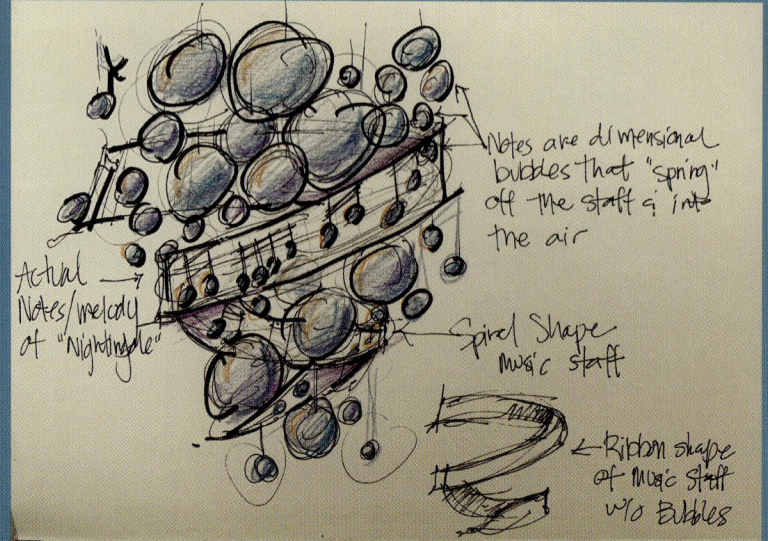

But we're not taking you into the movie world," Weiss noted. "We are taking you into this fantasy environment. Cinderella is there and it's inspired by magic and her wish coming true."

As for the ship from Norway, Weiss said, "the Keg & Compass [a pub on Deck 5] is another terrific example of an inspired motif coming into play, where it's not quite a totally immersive time and place story—"[since] we're not taking you back into 1850s Norway," Weiss emphasized. "But we're trying to give you that feeling of being in a different place and having a completely authentic experience without being absolutely architecturally authentic.

"I think of it as architectural cosplay," she added. "It reminds you of the place, it feels like the story, it has the colors and evokes the emotions, but it isn't specific. Even if you can't put your finger on it, it still feels like a Disney space. The Cove Café [the adults-only lounge on Deck 13] is an example of that. It hints at *Moana*, but it isn't exactly *Moana*."

As to how they split the work, Handke explained, "Claire [Weiss] gets to do what she does best, bring storytelling in her unique way to the motif areas. I get to focus on the more immersive spaces, the places that are more story-driven and have more of an emphasis on the actual movies."

Determining what stories to tell and where to tell them is not an easy process, according to Pam Rawlins. "A lot more thought goes into this than just picking from a list and saying, '*Inside Out* is our sweetshop theme.' What do the characters represent? How are they represented? How does it help tell the story?

"And then there's this checklist we have to do," she continued. "Say we introduce a Pixar-themed space like Wheezy's Freezies [based on the character from *Toy Story*] on the top deck. We come up with a concept and we may like it, but we also need buy-in from Disney Cruise Line; and marketing and public relations are also totally involved.

"Then we have to make sure Pixar approves how we're using the characters," Rawlins cautioned. "Also included are additional legalities and clearance rights. [So] depending on what the property is, we're dealing with Walt Disney Animation Studios, Marvel, *Star Wars*."

And remember, this was just for the *Disney Wish*, the first of three ships that the team was designing in rapid succession.

"We're changing each ship so drastically in terms of the themes and motifs," said Rawlins. "It certainly disrupts our norm in the way we've designed our ships in the past, but I think it's really welcome—and a real challenge."

In the end, though, Rawlins knows all that effort is worth it: "I believe our guests are going to be extremely surprised when they walk the ship and see these spaces."

Pam Rawlins

Executive Producer, Walt Disney Imagineering

It takes talent, skill, and passion to become an Imaginer. And, as any Imagineer will tell you, it often takes a bit of luck.

Say hello to Pam Rawlins, a West Virginia native who has taken her talents, skills, and passions—and generated her own luck.

It started when she was studying communications at West Virginia University.

"Disney put fliers up in the Student Union that they were coming to interview for the Walt Disney World College Program," she said. "I had a sorority sister who had been part of the program and was really excited about it. She told me that everybody who applied was putting in as an Attractions host. I thought, 'Well, I'll put Food & Beverage first.'

"I believe there were 1,400 people who showed up for the interview," she continued, "but there were only seven of us who were taken. From what I understand, my choice probably cinched my spot as one of the seven."

Pam ended up working at food carts in the Magic Kingdom.

"I did that all summer, really enjoyed it," she said, "and when I went back for my last year at WVU, I knew that Disney was going to be my career."

So what did she do next? She blindly called Walt Disney World Resort Information.

"I got someone who was probably very much like me at the time and they went, 'Guest Relations is where you want to work. You get your foot in the door.'"

As luck would (again) have it, Disney–MGM Studios (now Disney's Hollywood Studios) had just opened and "they needed Guest Relations people to replace the cast members that had been moved from EPCOT to the new park," said Pam. "Because I had talked with this woman on the phone, she told me everything to say and what to do, so the recruiter was super impressed and soon I was at EPCOT Guest Relations as a tour guide."

(It should be mentioned here, just in passing, that Pam's husband was also a tour guide at EPCOT. "I trained him and then we started dating," she said.)

Finally, we come to Walt Disney Imagineering.

"I heard from a co-worker whose husband worked at Imagineering that Imagineering was looking for people with operations experience to work with Imagineers who were building Disney's Animal Kingdom," she said. "I was asked if I would be interested in interviewing.

"Of course I said yes," Pam added. "I was hired in January 1996 and I've never looked back."

Echoing Headquarters, the control center inside Riley's mind in the Pixar Animation Studios film *Inside Out*, the area Inside Out: Joyful Sweets emphasizes the sweeter side of the hockey-loving eleven-year-old girl by way of this dessert shop featuring her five emotions—Joy, Sadness, Anger, Fear, and Disgust.

A large section of the *Disney Wish* known as the gigablock is docked at Meyer Werft, the massive shipyard in the northern Emsland region of Germany that is building three new ships for the Disney Cruise Line.

It Takes a (Worldwide) Village to Raise a Ship

As the *Disney Wish* progressed from concept and design to construction and installation, the epicenter of activity shifted from the Walt Disney Imagineering offices in Glendale, California, and Celebration, Florida, to the small village of Papenburg, Germany, where, apparently, everybody knows your name. "I went to a restaurant with some of my colleagues, and [with it] being such a small town the waiter who was there knew me from before," said Bob Girardi, principal show manager, Walt Disney Imagineering, who, a decade previously, had worked on the *Disney Dream* and the *Disney Fantasy* in Papenburg. "The owner said, 'Hey, welcome back. Good to see you again.' My colleagues said, 'Are you the mayor of this town?'"

While there is a mayor of Papenburg, it's not hard to see how Girardi could be mistaken for a local bigwig. Shipbuilding rules in this hamlet of about thirty-eight thousand located on the river Ems in Lower Saxony in the northwest part of the country. It is home to 225-year-old Meyer Werft, one of the oldest, largest, and most modern shipyards in the world.

The Disney team in Germany is made up largely of people hired locally. Many are shipbuilding veterans, who had previously worked on the *Disney Dream* and the *Disney Fantasy*.

More than a few Imagineers, however, have relocated to Germany. Some, like Bob Tracht and Bob Girardi, have worked on Disney ships for years; others—including Tim Hall, Lauren Fong, and Kyle and Alyssa Bilot—are new to the experience.

"It's kind of like summer camp," said Hall, project design manager, Walt Disney Imagineering, "where even though we live in different houses, we have these shared experiences. Most other projects I've worked on, you go home at the end of the day, and you spend it with your other friends and family, but here we're all family at night and on the weekends, too, because that's who's here. It's a special experience and a special bond."

For Fong, assistant project manager, Walt Disney Imagineering, on AquaMouse, Disney's Oceaneer Club, and several other areas, it's a homecoming of sorts. She participated in an exchange program for young professionals after graduating from college in the United States, and after the yearlong internship ended, she did what you're not supposed to do: "I stayed in Germany."

That decision did give her the opportunity to learn the language, make some friends, and go to graduate school in Europe.

"If you had told me when I was a senior at Columbia [University in New York] that I was going to move to Germany and learn German and then move back there again [to work], I would have said you're crazy," said Fong.

For the Bilots, it seemed crazy as well—to end up in Papenburg.

"We had both just bought brand-new cars," said Kyle Bilot, assistant project manager, Walt Disney Imagineering. "We had built a house from scratch only three years before that; we had a one-year-old boy, and then it was, 'Will you move to Germany for an indefinite period of time?' It was a big question mark of if we would do it. We went back and forth, evaluated a lot. By this time, I had already been making trips to Germany, so I knew about Papenburg, although, frankly, it made me wonder how I could live there that many years after growing up in a big city [Milwaukee] and then living in Orlando."

"This [Papenburg] is actually a big city compared to where I'm from," observed Alyssa Bilot, who grew up on a farm in Wisconsin and is also an assistant manager with Walt Disney Imagineering. "But I always knew we had to take the opportunity. He [Kyle] says we went back and forth, but I knew we had to do it if presented to us. The opportunity to work abroad anywhere, you just have to grab it when you get the chance. He was, like, 'Umm,' and I was, like, 'Yup!'"

The Bilots' situation has provided them with a unique perspective on their work-life balance. "[With] us being married and working on the same project," said Kyle, "you hear the same issues at work and the other person always knows what's going on; that's the challenge we have more than anybody. You're surrounded by it all the time."

"Work and life have melded together, and it just becomes . . . life," added Alyssa. "I'm okay with that because I'm happy doing what I'm doing."

"We do have date nights where we make it a rule that we're not going to talk about kids or cruise ships . . . ," said Kyle.

". . . and then we have nothing to talk about," interjected Alyssa.

Another Imagineering expatriate was Claire Weiss, who had the unique experience of literally moving with

(Near right) Meghan Moore, Monica Ireland, James Willoughby, David Atwood, Frans Nellen, Giacomo Panicucci, Bob Tracht, and Steve Read (sitting) gather for the signing of the contract that finalized the basic design of the *Disney Wish*.

"Hello! Welcome to Papenburg," reads the sign in the small German city where the *Disney Wish* was built (far right); the offices of Walt Disney Imagineering Germany and Disney Cruise Line at Meyer Werft (bottom left); the "mayor" of Papenburg, Bob Girardi (second from left), holds court at a local restaurant with (from left) Anton Erasmus of YSA Design, Meghan Moore, lighting consultant Paul Scott and David Atwood (bottom right).

the project as it went from concept in California, to design in Florida, to construction in Germany.

"It allowed me to get to know each team as the work shifted to each place," she said. "That was really key because those relationships have stretched into this final segment of the project and now half of the team is here with me in Germany."

Unlike most of her colleagues, though, Weiss decided not to live in Papenburg, opting for Oldenburg, about an hour by car from Meyer Werft.

"I've developed a limited group of German friends just by walking my dog," she said. "It's actually allowed me to feel much more a part of my neighborhood, except that I think my dog has more friends than I do."

Despite that physical distance from her colleagues, Weiss does manage to stay in touch . . . maybe even a little more than she'd like. "Tim [Hall] likes to call, and when I don't answer he just continues to call," she said. "Although one time he said, 'Fire!' . . . and I said, 'Tim, everything can't be a fire. I'm in a meeting.' And he said, 'No, no, there's actually a fire in my apartment and I need help.'"

Prior to joining the *Disney Wish* creative team, Weiss had worked on other projects for Walt Disney Imagineering, though those were firmly situated on terra firma.

"It's just a completely different way of designing," Weiss said of working on a cruise ship. "Most of what you do on land, you can't do on a cruise ship for all kinds of reasons: weight, flammability, cost.

"What's cool about it is that we can rely on our design partners," she noted. "They know the tools to use, and our job is to be the shepherds of the story. I couldn't tell you which laminates look exactly like stone—well, I can now—but at the beginning I didn't even know you could have a stone-looking laminate. My job is to look at the experience, and their job is to tell us how to get there."

It didn't start out this way for the Imagineers.

"With the *Disney Magic*, we learned how important consultants are to the process," explained Bob Tracht, executive cruise ship design and program management director, Walt Disney Imagineering. "They know the standards and the materials you're limited to on a cruise ship because of all the regulations."

But at the same time, those consultants have learned that Disney tries to get the most out of those limitations.

"When I talk to vendors, design partners, and outfitters for Disney Cruise Line," said Bob Girardi, "one of the very first things I always tell them is that the thing you have to remember about us is that we are not a cruise line company that happens to have some entertainment on board.

Labels such as these two (top right and left) identify every part, piece, and parcel of the *Disney Wish*, from steel beams and drywall to shower stalls and stateroom sofas. The 0705 refers to the ship's project number at the shipyard.

Ready to build (bottom): Kat Lashley (left) and Marisa Maynard at the Walt Disney Imagineering offices at Celebration, Florida, with hard hats for the Imagineers who will be working on the *Disney Wish* in Germany.

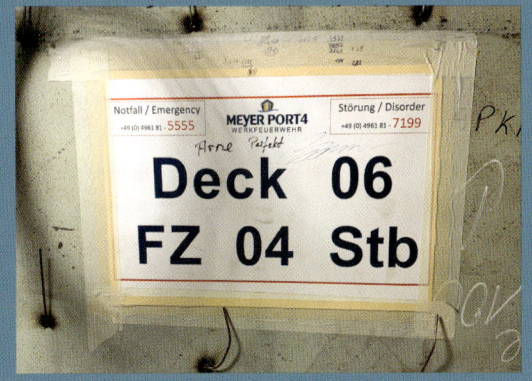

Imagineers traveled far and wide for inspiration while designing the *Disney Wish*. Alexis Cummins at Tillberg Design in Sweden (top); there's nothing more beautiful than springtime in Papenburg (bottom left); as members of the Disney team will attest to, including (bottom right, left to right) Lauren Fong, Jason Roberts, Manny Ramirez, James Willoughby, Danielle Duffy, Laura, Cabo, Alyssa Bilot, Kyle Bilot, Monica Ireland and (in front) Grayson and Calum Bilot.

We are an entertainment company that owns a cruise line. These are two very, very different things. We do not use technology for the sake of using technology, just because it's a cool wiggle light or a cool whatever. If it's not helping me tell the story to the guests, then we don't want to see it and we don't want to use it."

Jan Krefting, senior architect/partner at YSA Design in Oslo, Norway, one of the design partners on the *Disney Wish*, knew all too well about collaborating with Disney.

"I had been told that working with Walt Disney Imagineering was fun and challenging," he said, "but [that] it would also be frustrating."

He remembered a time when, after his team had already made several proposals that they thought had hit their mark and had outright found the perfect solution, the reaction from Disney wasn't what he hoped for.

"'Yes, that was nice,' was the response," said Krefting. "'But maybe you could do a little twist and present again?' followed.

"By this time, we were ready to pull our gray hair out," he added, "but we did it and, even if it was hard to admit, the final result was better."

At the end of the day, this is an incredibly complex project that Disney and its Imagineers could not do alone.

"That's why we have this whole global team," said Claire Weiss. "All of these people are the experts in the industry. They tell us what we can use, where we can use it, what we can't use, and what all the special ship things we need to do are. They do that part, and we tell the story."

As the pages on the calendar got flipped over and reached into 2020, anticipation was building across the board. The design firms were finalizing their drawings; mock-ups and production were taking place in Florida, California, Germany, and other parts of the world; and construction was already underway on the Floating Engine Room Unit (the FERU) at Neptun Werft, Meyer Werft's sister shipyard in Rostock, Germany. Other various elements/parts of *Disney Wish* were also already being built in Gdansk in Poland, and Klaipeda, Lithuania.

"This is my favorite part of any project," said David Atwood. "To me, as fun as design is and as amazing as it is to see a concept hand sketch go to a full set of construction documents, I love the building process, so getting to this phase of the project, it's just phenomenal, such an exciting thing to see."

There was just one slight hiccup.

"For about two years, we had our beats down," said Pam Rawlins, the executive producer. "People were traveling, we had major milestones we were accomplishing with the shipyard. We were rocking. We had our marathon, we had our pace, we were breathing correctly, everything is good—and then the pandemic hit."

Bob Girardi

Show Manager Principal, Walt Disney Imagineering

It was 1990 and Bob Girardi was plying his trade as an automated lighting specialist in and around the Tampa area, where he grew up.

He'd worked in local theaters, at Busch Gardens Tampa, and for a number of rock 'n' roll acts that had come through town, including Kiss, Scorpions, and one called Ice Nine and the Belching Penguins.

And then someone from the Walt Disney World Resort called.

"'Can you help us get this straightened out from a technical standpoint?'" he said he was asked of the Mannequins nightclub at the recently opened Pleasure Island. "I said, 'Sure, I'll come in and help you out and stay for a year and then go back out on the road.' More than thirty years later, here I am, still here."

It's not surprising to see why he stayed. After solving the issues at Mannequins, he started working at the West End Stage at the Pleasure Island, which today has been absorbed into Disney Springs.

"We had a large number of huge acts that went through there," he said, "and I have a stack of credentials to show for it.

"We brought on bands that would later become huge stars like Sister Hazel, Hootie & the Blowfish, Sheryl Crow," he added. "I ran lights for N'SYNC with Justin Timberlake, and we also had Britney Spears and the Backstreet Boys and all those guys."

His work eventually caught the attention of Walt Disney Imagineering, which was looking for a show and production manager on the two Disney ships, the *Disney Dream* and the *Disney Fantasy*, that were about to start construction.

"I had just got married," said Bob, "I was in the airport on my way out to my honeymoon and I got a call from someone named Bob Tracht. He said, 'Listen, we need some help here.' So, after the honeymoon, I transferred to Imagineering."

Bob has always been a big fan of Disney, dating back to his childhood.

"We made one trip per year to Walt Disney World," he said. "I just remember that long drive up I-4. I remember how the median looked as we got closer. It's got very tall trees in it and I would just get so excited because I knew we were close."

Even today, he remains mesmerized by the magic.

"When I was working out of the Imagineering office at Epcot, if I had had a particularly bad day, you'd probably find me following the character bus around the World Showcase promenade," he said. "If you want to get pixie-dusted, that's the place. You go out there and follow the bus and just watch the characters meet guests. That's what it's all about."

Tim Hall

Project Design Manager, Walt Disney Imagineering

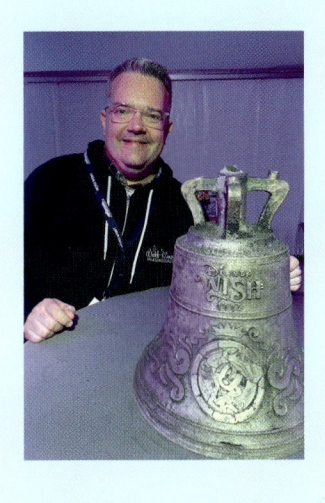

As he approached forty, after years of managing the design and construction of large-scale Caribbean resort projects, Tim Hall decided to go back to school, at the University of Florida.

Sure, part of it was that he wanted a master's degree in construction management (he already had a degree in architecture from the University of Cincinnati), but he also had an ulterior motive.

"It meant I could qualify for an internship at Disney," he said.

It had been a lifelong dream of Tim's to work at Walt Disney Imagineering. He had gotten a taste of it as part of the University of Cincinnati's cooperative education program, working with the Imagineering master planning studio in Celebration, Florida, for six months.

"Up to that time, it was the coolest job I ever had," he said. "After I went back to school to finish my architecture degree, I kept trying to get back to Disney. It just took me twenty years."

The dream started in Stow, Ohio, in the Akron area.

"I remember watching the Easter and Christmas Parades and the 'making of' specials that were on TV," said Tim, "They always had these 'this is the attraction that is coming' or 'this is a hotel we're building, and this is how we make it' features. Somewhere in my brain, I put it together, 'Oh, people build this stuff and there must be a group of people that actually designs it as well,' so that's what got my interest in being an architect and becoming an Imagineer."

It also helped that his dad had his own construction company, which means he was always around construction projects, and, perhaps more importantly, his family vacationed at the Walt Disney World Resort every year.

"We had a camper and stayed at Fort Wilderness," he said. "We would drive down for at least a week, if not more, every year."

After earning that master's degree at the University of Florida, he did land a position at Disney. "I ended up getting a full-time job as a scheduler in what's called Facility Asset Management, working at the Magic Kingdom," he said, "always keeping my eye on Imagineering. I gave myself a year to become an Imagineer."

And nine months after starting at Walt Disney World, that's exactly what happened. Tim was tapped to join Imagineering as a design manager, eventually joining the *Disney Wish* project team.

I love Disney and Imagineering," he said, "because we are working on these massively creative and technologically intensive projects that our teams figure how to design, build, and operate. That's what gets me excited."

With the lockdown in full swing, Laura Cabo used her garage as a makeshift design studio and conference room, doing virtual meetings day and night.

Lockdown, *Schmockdown*

"I remember very well it was a Wednesday in mid-March [2020]," said Philip Gennotte. "I told our team here in Germany that you should pack what you need because we may have to work from home, and that Friday that's exactly what happened." It's tough enough to design and build a project as big and complex as an immense ocean liner under the best of circumstances, but throw in a pandemic and you have the makings of a catastrophe!

And that's exactly what Gennotte feared. "I thought it was going to be a disaster," he said of lockdowns that took place around the world, including in Germany, "because I have a team here that I'm responsible for, a team where many of them don't speak the language, they just moved here, and they're on their own in their apartments.

"I put a lot effort into reaching out to them, making sure they were safe and okay," recounted Gennotte. "That their Internet was working okay; do they need anything?"

"We thought it was just going to be for a couple of weeks," said David Atwood, "and then maybe until June. And then it was the end of the year. And then it just kept going on and on and on."

And yet, despite a few delays, primarily at the shipyard and the other production facilities—which had to figure out procedures and protocols for moving forward safely—the project proceeded with barely a hitch.

"I don't want to minimize how tough it was on everybody," said Atwood, "but I think because of the way the project was set up—we had design firms in Norway,

Sweden, and England, and team members in California, Florida, and at the shipyard here in Germany—we were already from very early in the project collaborating remotely. We just went from conference rooms and desks to our homes."

Added Gennotte: "We'd always done meetings by Zoom just because we have team members and design partners all over the world. But before [the pandemic erupted] you'd have a conference room filled with people and one or two other people coming into the conference room on screens. Those people would always feel a little left out.

"But now everyone is on equal footing. Oddly enough, it really helped bring everyone together in a different, more connected way," he believed.

And, as is human nature, everyone was pitching in every way they could to make sure the work continued.

"The logistics have been challenging," said Bob Tracht. "Imagineers have turned their own homes into offices, mock-up spaces, and production facilities. It's been tough enough continuing the design process remotely, but then you add in mock-ups and construction, and it gets very complicated. But through all this, the health and safety of everyone involved was a top priority.

"The fact that we've been able to do this is just a tribute to Imagineers and their dedication," he added.

Claire Weiss was one of those Imagineers on the front lines.

"The pandemic forced people to jump in and help out where others couldn't physically be," she said. "We still went to mock-ups. We still had to meet with vendors and go to production facilities and warehouses. We still had to build the ship. But we were always taking the proper precautions and following every safety measure in place, including wearing face coverings and keeping the proper distance.

"This has been a spectacular feat of teamwork," Weiss continued. "We've grown to lean on each other more than we normally would have.

"On the other hand," she added, "I never knew how nice my apartment was until I spent most of a year sitting in it."

It wasn't just in Germany where team members were making adjustments.

"[Here in Florida] we're a small, tight team, and I think if anything the pandemic brought us closer," said Stacie Covington, producer, Walt Disney Imagineering. "The first couple months were hard, but after that we developed a rhythm where it was easier because we were always accessible to one another."

Covington further noted, "Even when we were in the office together, there were a lot of times when, because of meetings or vendor visits or driving somewhere, we were missing each other. But now every day we had a connection. We knew where each other was. Everyone was home and easily reachable by Zoom. I think it actually made us a little more efficient, though

Working from home presented all manner of challenges—and opportunities—for the *Disney Wish* design and delivery team. Virtual tours of the ever-evolving ship were offered over Zoom (bottom right); Danielle Duffy was able to spread carpet samples all over a living room (and work in shorts!) (bottom left); Philip Gennotte (top right) led weekly meetings—virtually—including a holiday party in 2020 (center right); and even pets got into the act with Katrina Alvarez's dog Enzo (top left) checking out a carpet weave, while Lina Barr's golden retriever Lily (top center) helped with color samples.

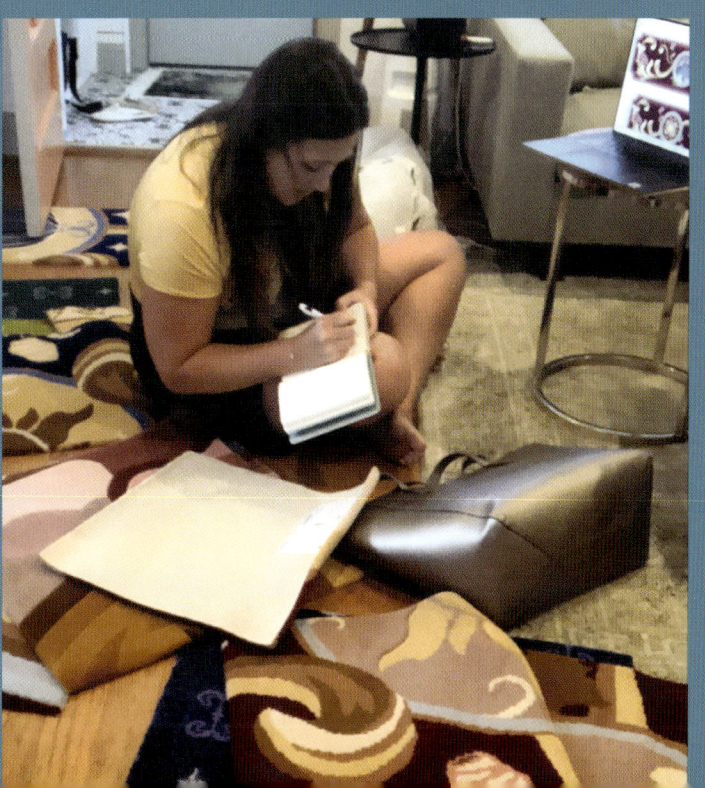

With health and safety protocols in place, construction moved forward on the *Disney Wish*. Block 1 (top) is craned onto the FERU (Floating Engine Room Unit), while more blocks stand ready to be lifted into place (center right); the FERU (center left) is floated into the hall as blocks wait to be lifted onto it (center right); Kyle Bilot (left) and David Atwood in front of an elevator block, the first to go on the ship (bottom left); in the bitter cold of Rasdorf, Germany (bottom right), (left to right) Luca Graffigna, Doug Larsen, Zvonimir Vidak, Lauren Fong, Luciano Falcone, and Markus Rauna gather to conduct a FAT (factory acceptance test) on AquaMouse conveyor equipment.

nothing beats working with the team in person."

Even before the lockdowns Gennotte had made an effort to keep his far-flung team informed and connected, hosting bimonthly staff meetings via Zoom that featured design and construction updates, presentations from team members on what they were doing, and celebrations of birthdays and work anniversaries.

"We didn't do this on the previous ships," said Gennotte. "We've been able to give [to] the team more than ever before, to keep them connected to the progress of the ship. To me, that is the biggest thing we've been able to do."

The meetings were a logistical challenge in and of themselves, since a 4:00 p.m. start in Germany meant it was 10:00 a.m. in Florida, 7:00 a.m. in California, and for poor systems engineer Connie Tang (before she moved to Germany) 10:00 p.m. in Hong Kong.

And those gatherings became even more crucial after the world essentially went into shutdown mode.

To that end, if anything, Gennotte upped the ante, conducting team outreach meetings to discuss what was happening with the *Disney Wish* as a major rapidly spreading health crisis was unfolding worldwide and to check in on team members. Zoom meetings went beyond business, with Gennotte holding seasonal celebrations, commemorating big events in the ship's progress, and providing "virtual" tours of the ship once construction started.

"We've tried really hard to involve everyone on the rest of the team in the process here," said Tim Hall, project design manager, Walt Disney Imagineering. "For the remote people, they're probably getting more exposure to the process because of the pandemic. We probably never would have done walk-through videos of the ship or video reviews of mock-ups."

Updates were a constant. "Part of my role was getting everyone comfortable with doing business this way," he said. "Plus, I wanted to do virtual tours of the ship being built. The yard wasn't used to that. They were

concerned about their trade secrets getting out if someone was bringing a camera into their facilities. I convinced them [to take a chance]. I said, 'This is the situation we have. We have to do this.' And so we set up rules on paper with them on how we do this. We even had the yard manager join us so he could see why and how we were doing this. It's amazing what the phones can do now, the quality and everything."

Gennotte encouraged individual teams to host their own events—and most took him up on it.

"I set up some virtual socializing for my team," said Bob Girardi. "We did it twice a week. We called them 'Happy Half-Hours.' We'd all get on the call, and we weren't allowed to talk about work. We'd do trivia or talk about that lawyer that was a cat and couldn't turn his cat-face filter off.

"We had to do that because we still had to come together as a team," insisted Girardi.

Pam Rawlins put together a virtual social hour every other week with the

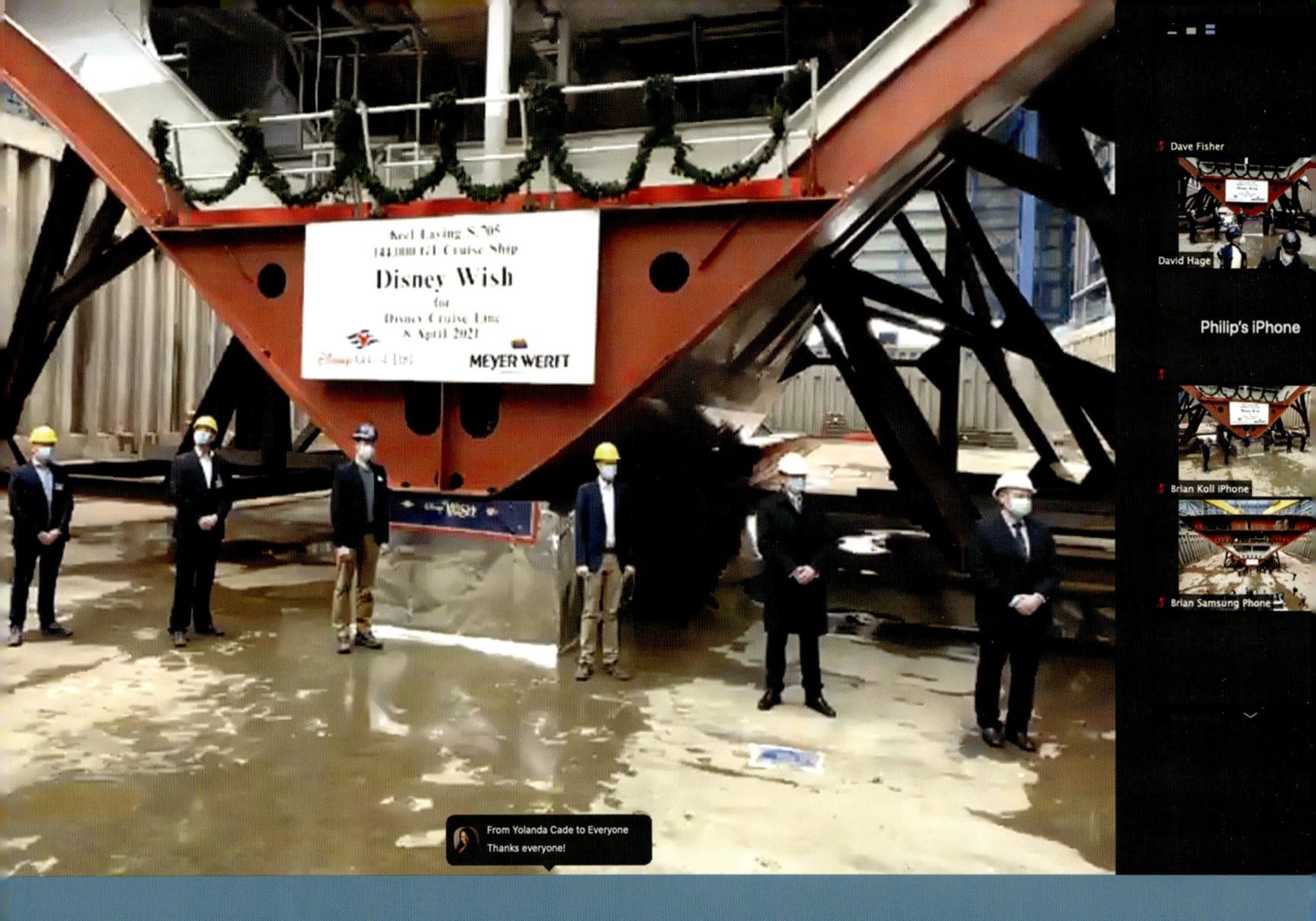

A huge milestone in the building of the *Disney Wish*: the laying of the keel—a ceremony held virtually over Zoom, of course.

development and production team that featured increasingly elaborate trivia games. Each member of the group was seemingly trying to outdo the one who had put together a game preceding theirs.

And, like many people stuck at home during the pandemic, Imagineers cultivated a lot of hobbies: gardening, baking bread, doing jigsaw puzzles, knitting scarves, adopting pets, watching Disney+ from start to finish, painting and drawing, writing this book, and having babies.

Wait, what was that last one again?

Imagineers produced no fewer than a dozen offspring during the course of the project. In case you're wondering, that's about 10 percent of the team.

Among those welcoming a child were Alyssa and Kyle Bilot, whose son Calum joined older brother Grayson in November 2020.

"I found out I was pregnant a week after we were told we had to work from home," said Alyssa. "It was really hard because we don't have family here [in Germany], but we have people who help us out, who were really kind and offered their time to watch the kids when we needed them to."

"We really had a network we could depend on," added Kyle.

It was just another example of Imagineers watching out for each other.

"The resilience of the team has been amazing," said Bob Girardi. "You can't tell them they can't do anything because they figure out a way to do it."

"You *never* say never," said Gennotte, "but I can't imagine being able to build ships like these in a pandemic even ten years ago. Today I keep wondering, is this even possible? How do we do this? I'm pinching myself. This is incredible."

"We have completely revolutionized how we work together," he observed. "We now actually do more together than ever before. We are so connected and so well informed and everything just seems to be more efficient."

David Atwood summed it up this way: "In some ways, it was fairly seamless because we were all so accustomed to it anyway. From a work standpoint, it wasn't that bad early on. It was tough not seeing your friends and colleagues in the office every day. So much can happen when you have those watercooler conversations or [by] just popping your head into someone's office. But not being able to travel and have the design firms and your colleagues in person and looking over a set of drawings, it took a little while to be efficient. [Still,] all in all, it went pretty well."

But then came the main event.

"It did get harder when we actually started building the ship," Atwood continued. "It would have been great to go out in the morning and walk the ship and work from the office in the afternoon. The biggest pain was just the isolation, but the work itself, it was pretty much full speed ahead."

Kyle and Alyssa Bilot

Assistant Project Managers, Walt Disney Imagineering

Kyle and Alyssa Bilot both grew up in Wisconsin, she a small-town farm girl, he a big city boy from Milwaukee.

They both studied architecture at the University of Wisconsin, Milwaukee, but even then they didn't meet until she was a sophomore and he was a junior.

"We met through mutual friends at an American Institute of Architecture students conference in Denver, Colorado," said Kyle. "So we didn't even meet in the city where our school was. We met on a school trip in a different state."

Kyle really didn't know what he wanted to do after graduation, so he went to graduate school. "He's got the master's," said Alyssa, "he won't let you forget that."

As for Alyssa, she knew exactly what she going to do and had for many years.

"I knew on my first trip to Walt Disney World that I wanted to work there," she said. "I was four and my mom remembers me saying, 'Someday I want to make magic there.'"

Working at Walt Disney Imagineering became a particular obsession for Alyssa. She read the big coffee table book *Walt Disney Imagineering: A Behind the Dreams Look at Making the Magic Real*. She did the "Dine with an Imagineer" program at Disney's Hollywood Studios—twice. She participated in the Disney College Program.

After she started dating Kyle, she made things very clear to him.

"'I'm going to work for Disney someday,'" she said she told him, "'so you better deal with that.' He knew up front that was what was going to happen in life."

And happen it did.

"After the second interview, when it looked like I was going to get the job at Imagineering in Florida," she said, "I had that conversation with Kyle. 'Okay, Kyle, you always talked about not having to make a decision until we reached the bridge. Well, the bridge is here. What are you going to do?'"

Of course, this has a happy ending. Kyle joined Alyssa in Florida in 2012, a few months after she took the job at Walt Disney Imagineering and, soon after, he ended up there as well. The next bridge? Marriage, in June 2014.

Today, they are living in Germany, assistant project managers on the *Disney Wish* (and the two ships to follow), responsible for many of the restaurants, lounges, shops, and other public and crew spaces in the vicinity of the Grand Hall.

"It's called the Bilot part of the ships," said Kyle.

Laura Cabo was eventually able to open her garage for meetings with the California contingent of the design team, (left to right) Christine Dantzler, Cabo, Davey Feder, Dave Fisher, Lina Barr, and Megan Moore (not to be confused with Meghan Moore, on the *Disney Wish* team in Germany).

A worker applies the finished touches to Captain Minnie on the bow of the *Disney Wish*.

Full Speed Ahead

June 17, 2020, was all set to be a momentous day (among many momentous days) for the *Disney Wish* and everyone working on it. "The first steel cutting for the ship is a truly significant milestone," said Denise Case, director, Creative Entertainment, Disney Cruise Line, "and we were going to celebrate that in a big way. We had sent a scenic package to Papenburg [Germany] for an event we were going to stage—and Covid hit. It never happened."

Instead, a small team of Imagineers watched as steel production began on Block 1, the first of eighty-five blocks to be built at Meyer Werft in Papenburg. (Another sixteen "standard" blocks were built in Poland and Lithuania, and then transported to Papenburg.) That first block encompassed part of the lobby for the aft elevators, as well as a portion of Disney's Oceaneer Club.

The acknowledgement of this noteworthy event was low-key and understated, but it did represent a major breakthrough for both Disney and Meyer Werft. It meant that, thanks to the establishment of strict health and safety protocols at the shipyard, including the wearing of face coverings and implementing of social distancing rules, the construction of the ship was able to move forward.

Disney Wish would be the only ship to be built at Meyer Werft during the pandemic.

While the first steel for the ship was being cut in Papenburg, 260 miles to the northeast in Rostock, Germany, two extremely large liquid natural gas tanks were about to join the ship's five engines on the FERU. "Photos can't capture how immense the inside of these are," according to Philip Gennotte. Work crews, however, were undaunted, and later in the year, the entire unit, would become the base of the "gigablock," the central part of the ship upon which would be stacked more than forty more blocks, including Block 1. These blocks would make up the crew quarters, guest staterooms, the main hall, the pools, and many other public spaces.

With more than four years of planning behind it, and a little more than a year till launch, the sprint to the *Disney Wish*'s finish began on January 8, 2021, when the first block (Block #1, of course) was lifted onto the FERU. Over the next several months, the ship quickly began to take shape, with blocks completed and lifted into place practically every day.

"It's just amazing to see the construction of the ship in the hall," said David Atwood, project manager principal, Walt Disney Imagineering. "They've got this massive overhead crane that is lifting a block that's the entire width of the ship and several decks high. It's an incredible process. It's amazing to see how quickly the ship goes together."

The last block was lifted onto the gigablock in March 2021 and, later that month, specifically March 26, the entire structure was floated out of Hall 6 at Meyer Werft in the wee hours of the morning.

It took two hours, yet the whole thing was seen over Zoom by team members around the world—and even by a few vloggers who somehow got advance word of the undertaking and were there to capture it for their Twitter feeds and websites.

The repositioning of the gigablock enabled construction on the other two parts of the ship (the foreship and the aft ship), and for the staging of another momentous undertaking: the laying of the keel.

This maritime tradition involves laying a coin (this one featuring 705, the hull number, on one side and Captain Minnie on the other) under the keel of the ship to celebrate the ceremonial beginning of the ship's construction (never mind that that had actually occurred almost a year before).

Normally, this would be a major ceremony, a press event filled with pomp and circumstance and Disney characters, but again, like the steel cutting event, it was scaled down to a few representatives from Disney Cruise Line, Walt Disney Imagineering, and Meyer Werft.

Still, there were more than a hundred team members watching on their computers and mobile devices as a block weighing more than 1.6 million pounds was lowered into place.

"I sure hope that coin is strong," said Philip Gennotte.

With the ship's structure quickly evolving, work continued on everything that was going to go onto and into *Disney Wish*, a process that included what are known as mock-ups.

"Mock-ups give the team the opportunity to see things before they go in production," said Bob Tracht. "And we really get into it, knocking the chairs over, trying to scratch them, testing their 'sit-ability.'"

One type of chair, in particular, proved to be a challenge. "No one could

With the gigablock in the background, Walt Disney Imagineering and Disney Cruise Line held a Shipyard Appreciation Day for all the workers at Meyer Werft who had labored through extraordinary circumstances to get the ship to this point.

The forward funnel arrives at Meyer Werft (top). This funnel houses the Wish Tower Suite and a passage, seen toward the rear of the structure, for AquaMouse; Rack City in Uplengen, Germany (bottom left), hosting more cables than Rapunzel has hair (bottom right).

actually sit on it," Tracht remembered. "It was more like a slide." Designers went back to the drawing board on that one.

For Tim Hall, mock-ups are also a chance to make sure things are being done the Disney way.

"I'm adamant about not seeing certain things, like speakers or light fixtures," he said. "They're a vital and essential part of the ship, but I don't want to see them. I want to see the magic that is created from those devices, not the actual devices themselves."

In all, there were nearly two hundred mock-ups for everything from chairs, cabinets, and towel and trash bins, to carpet, wallpaper, and complete rooms.

"Pre-pandemic, we would have flown in a handful of people, maybe eight or nine, from Florida and California," said Hall, who had staged more than a few of these demonstrations. "Now, I take photos ahead of time to send out to the group so everyone can review them before the meeting, set up the technology, run the camera, and facilitate the meeting. The downside is that not everyone is here live, but now I can have fifty people on a call reviewing the mock-up.

"When the pandemic had settled enough that people could travel again," he added, "we did bring them in for major reviews, but, because of the technology we had set up during the pandemic, we were still able to provide exposure and include many more people."

"It was the best way at the time until we could get back into the field and physically onto the ship," said Bob Girardi. "We made a huge amount of progress at a time when many projects were just trying to stay above water."

Another way Imagineers were able to make progress was by building a data center in Uplengen, Germany, about thirty miles from the shipyard in Papenburg.

"The thinking was that if we could get a facility and information technology could bring all of its racks there, we could set up the IT backbone for the entire ship in one space," said Girardi. "I could bring all one hundred-plus of my [entertainment] racks to that facility and we could bring all of the show systems online on the Disney Network in this warehouse. How cool would that be?

"So we built this venue in Uplengen," he said, "and we get our entertainment system ready to go. And then the pandemic hits! And we don't miss a beat. The fruits of our labor immediately pay off. We have an entire system that's on the corporate backbone. I can't physically walk to it because of the lockdown, but all of our engineers can pull it up as if it's already on the ship, which hasn't even been built yet, and work on it remotely."

So many rows were required to accommodate the 124 racks of components in the data center warehouse that a Rack City Road Map was created and posted on a wall near the entrance. "Streets" included Rack Back Power Road, Entertainment Road,

Wallstreet, and the always popular connecting "highway," Road to BBQ.

By doing it this way, the downtime on the *Disney Wish* was two weeks. (On the *Disney Dream* and the *Disney Fantasy*, the process took four and a half months).

Another time-saving innovation came in the construction of the guest staterooms and crew member cabins.

"The cabin production line is new to this class of ships," said Bob Tracht. "On the previous ships, we did traditional 'stick built' construction where hundreds if not thousands of pieces were carried to every cabin to produce in the actual location on the ship."

For the *Disney Wish*, an assembly line was set up at Ems PreCab, a manufacturing facility on the outskirts of Meyer Werft and just down the street from McDonald's (yes, even in Germany).

With a warehouse full of all the necessary materials and finished goods—walls, ceilings, carpets, cabling, lights, refrigerators, chairs, tables, beds, flat-screen TVs, safes, even the artwork to be hung on the walls—the rooms were assembled over the course of a twenty-two-step process.

"We're building thirty-five cabins a day, including and testing," said Tracht. "Everything is pre-cut and prepackaged; all of the cables are pre-bent before they get to the production line. They're pressure-tested before they go into the cabin. It's really something to see. It blows me away every time I see it."

Once this extensive process was completed, the cabins were trucked to the gigablock, then lifted and rolled into place right on the ship. Crews of outfitters came in later to hook up the plumbing and electrical, roll out the carpet, set up the furniture, hang the drapes, and apply the finishing touches. And, voila, just like that you had the largest, most comfortable guest staterooms found on any cruise ship.

By August 2021, the gigablock was ready to be moved back into Hall 6 and, in a feat of Herculean engineering and prodigious welding, it was joined with the foreship and the aft ship to form, once and for all, the *Disney Wish* as we know it and see it today. (Well, at least the shape of it. There was still plenty of painting and finishing to do.)

"When you work as hard as this team works," said Claire Weiss, "you sometimes forget what you're doing and who you're doing it for. You're so into the details and you're so into the specifics, and the second you're standing out on the FERU and they're craning in one of the funnels, you remember that you're building a giant ship.

"I don't think you ever get used to the scale and magnitude of this project," she added, "and I don't think we even completely wrap our minds around it. It really is this sense of specialness and this feeling that, *Wow, this thing we're working on truly is incredible.*"

How incredible? Let's find out by taking a look around the *Disney Wish*.

As more and more blocks were lifted into place, Imagineers were there to capture it, as Claire Weiss does here on her phone.

Alan Dunne

Manager, Technology, Disney Signature Experiences

When Alan Dunne was growing up in Lancashire, in the northwest part of England, he usually had only one thing on his mind.

"I was definitely into computers," he said. "When my mom bought us our first computer, I naturally took to programming and coding."

But at the same time, he developed an interest in theater. "In the part of the United Kingdom where I grew up, there were lots of shows," said Dunne. "We would go to a lot them. I started to wonder, even then, how I could bridge those two worlds."

It started in college, when the computer science and information systems major started working in a local theater during his spare time. "After graduation," he said, "I went in a completely different direction because, at the age of 22, the thought of coding for nine or 10 hours a day didn't really excite me. That's when I said, 'Okay, let's apply the programming knowledge that I have and apply it to entertainment.'"

The result was a stint working behind the scenes for a couple of seasons on shows in his hometown, followed by a job at a special effects company doing outdoor laser shows and indoor concerts. It's eventually what led him to Disney Cruise Line in 2003. Dunne's first assignment was as a show control engineer for a brand new production, "The Golden Mickey's," aboard the *Disney Wonder*.

And yet, he never strayed from his IT roots. After a year in entertainment, he transferred to the engineering department.

"The engineering department and entertainment technical were very different worlds," he said, "but by understanding both, I had the skill sets that gave me the opportunity to come off the *Disney Magic* and the *Disney Wonder* in 2009 to help build the *Disney Dream*.

Dunne has been building Disney ships ever since. Today, he's the manager of technology for the three new ships Disney Cruise Line is building, including the *Disney Wish*.

"Disney always pushes the envelope," he said. "They always are on the leading edge, pushing the industry forward, particularly in technology and entertainment technical. And that's where my two worlds, computer science and entertainment, have really come together."

The Disney Cruise Line also enabled him to create a third world: his family.

"I met my wife Ilona when she was working as a crew member on the *Disney Magic*," Alan said.

The two married in 2010 and have two daughters, Kaylin and Elisha, and a son, Aaron.

Giacomo Panicucci

Technical Superintendent, Walt Disney Imagineering

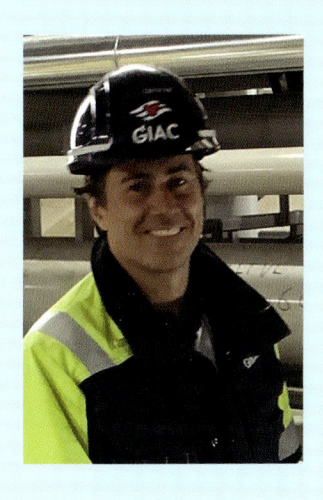

One day in summer, 1997, Giac Panicucci took a wrong turn in Venice, Italy, and it ended up completely changing his life. "I wasn't very familiar with the area," he said, "and, suddenly, I found myself at the back of this big, beautiful ship, which was under construction at that time."

He was intrigued. "It was the *Disney Magic*," he said. said Giac, who at that time was working for another cruise line company. "I didn't go right to Disney that day, but that became my goal, to one day work for the company."

That day came two years later when he joined the Disney Magic as a 3rd engineer. Eventually, he rose through the ranks to the role of Chief Engineer. Today, he's leading a team over seeing the design and installation of the machinery on the new Disney ships.

Pretty good for someone who didn't grow up anywhere close to the sea.

"I'm originally from Tuscany, Italy, close to Florence," Panicucci said. "It's a place quite far from the sea, so a career at sea was probably not in the planning."

But a friend of his father's, a ship captain, regaled him with stories of adventures to far-off places around the world—and he was hooked.

"I started to kind of get curious about it," he said, "and when it was time to for me to choose a career, I felt that I had already made my decision early in life, that I wanted to go to sea and that's how I started."

Panicucci attended the Italian Naval Academy and trained as an officer. After his service was over, he joined the Merchant Navy. By then, the cruise business was booming and, well, we've come full circle.

In the years since, he's continued to pursue his passion for the sea, but he's also found a home at Disney—and a little bit more.

During a tour of duty on the *Disney Wonder* in 2008, he met another crew member and today he, Ceci and their daughter Mia live in Orlando. Well, except when he's on a long-term assignment in Germany and they're all living there with him.

For Panicucci, the *Disney Wish* and the two that will follow offer opportunities to take advantage of continuing innovations in ship design, propulsion, engine and energy recovery systems that are making the new Disney ships not only the most beautiful on the high seas, but the most technologically advanced.

"It's not just about the new tools and techniques," he said. "It's always keeping in mind the two most important areas of concern. Ensuring and enhancing safety is one and the other is improving energy efficiencies because we have to be concerned about the environment."

Construction proceeds at a furious pace throughout 2021. The gigablock is floated out in the dead of night (top); the name is added (far left); and the ship truly begins to take shape (bottom right).

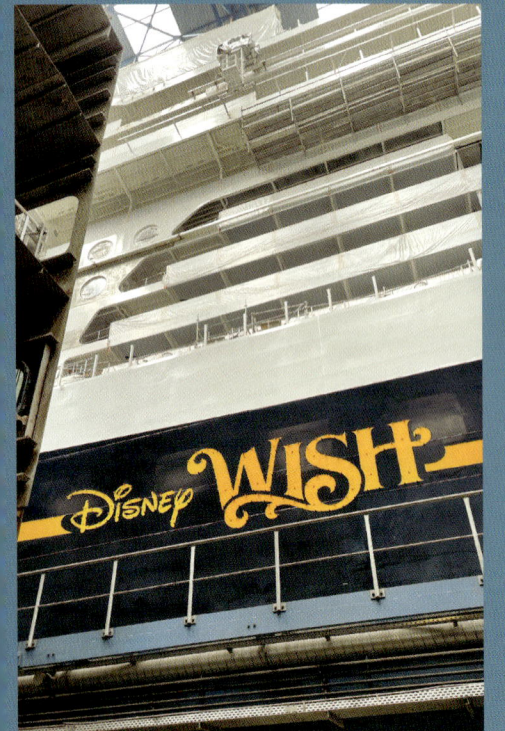

The aft funnel is lifted into place (top left); the gigablock returns, this time in the light of day (top right); and, soon, two halves will make a whole (bottom).

The bow, before and after. The filigree, including Captain Minnie, had already been etched into the surface, essentially enabling painters to paint by numbers.

Another big moment for the *Disney Wish* and its team: the block containing the ship's bridge is lifted into place.

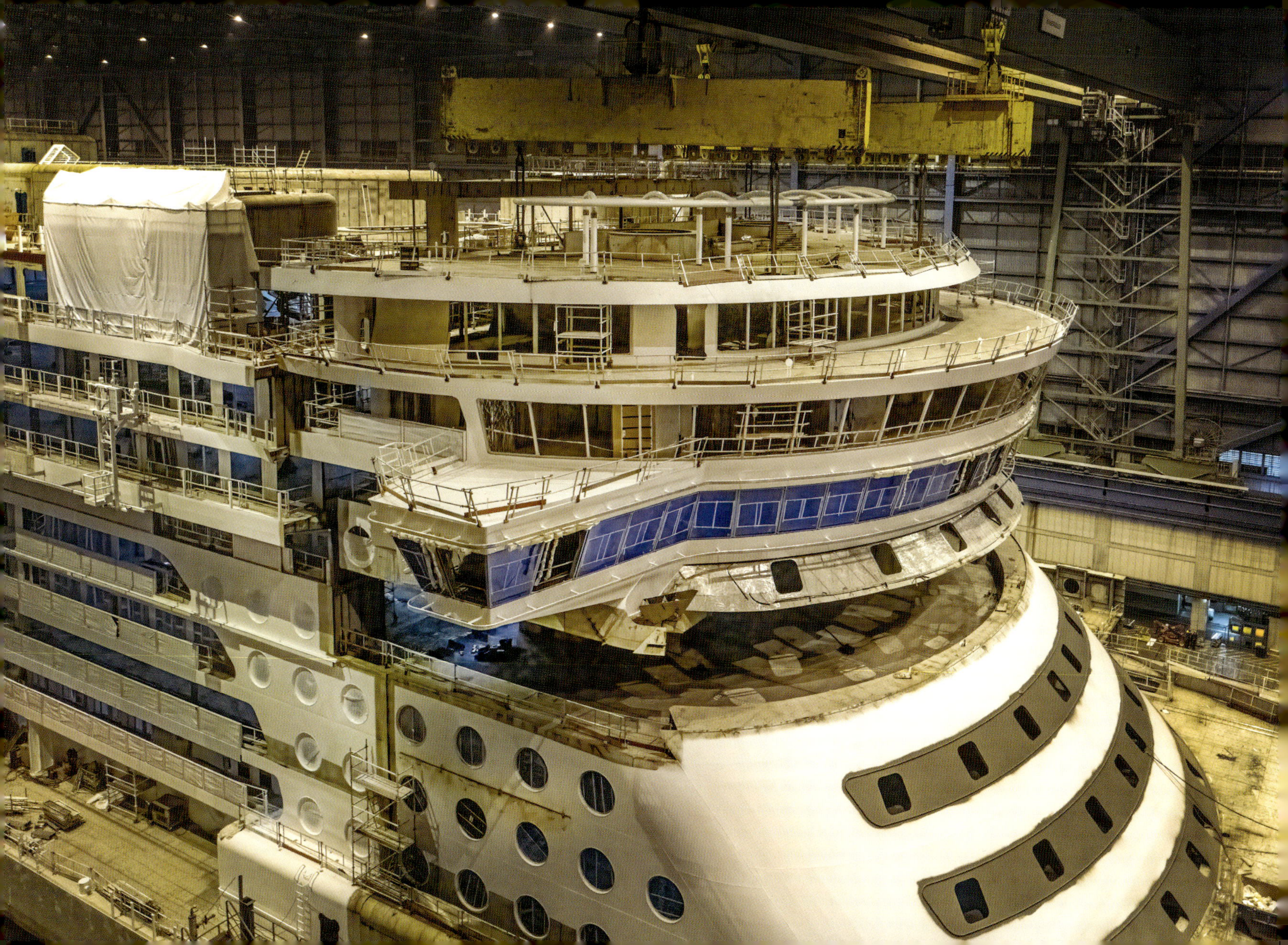

Zvonimir Vidak

Director, Marine Technical Operations, Walt Disney Imagineering

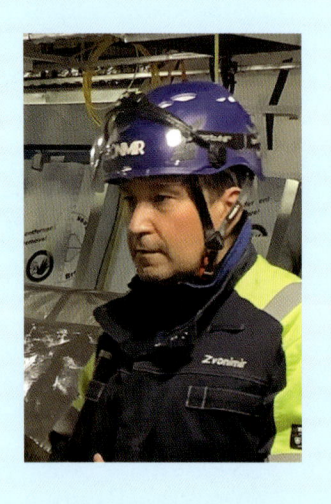

In 1998, Zvonimir Vidak, having just completed a ten-month commitment on a container ship, was sitting in Frankfurt Airport in Germany, waiting for a flight to Venice.

"I had long hair and a guitar," he said, "This group that was also flying to Italy, they approach me and one of them says, 'Are you one of us?'"

Zvonimir had no idea what they were talking about.

"They said, 'Well, are you a musician?' I said, 'No, this is just my hobby. I'm actually a sailor,' and they said, 'Well, we're sailors, too. We're going to join the *Disney Magic*.'

Zvonimir still had no clue.

"'What's the *Disney Magic*?' I asked, and they said it's the most beautiful ship on the seven seas. For me at the time, that didn't mean anything. But, you know, it's a small world after all and two years later I found myself working on the *Disney Magic* and, yes, it was the most beautiful ship on the seven seas."

By the way, the group he met at the Frankfurt Airport?

"They were actually the first cast for Walt Disney Theatre."

For Zvonimir, a life on the sea seemed not only certain, but inevitable. He was born and raised along the Adriatic Sea in Rijeka, Croatia, but, perhaps more importantly, his late father was a sea captain.

"I was very young when I got the idea that I would like to be a captain when I grew up and that never changed," he said.

After obtaining his shipmate's license and graduating university in Rijeka, Zvonimir began sailing on cargo ships. In 2000, he switched loads, swapping containers for passengers.

"During that time, I met someone with Disney Cruise Line—by this time, I knew what it was—and he said, 'Why don't you apply?' said Zvonimir. "So I did. I joined Disney Cruise Line in January 2001."

From there he worked his way up the ranks all the way to captain. And although he has been aboard all four of the previous Disney ships at one time or another, he has spent the most time on the *Disney Magic*, which led to one of his, ahem, most magical moments.

"I met my wife, Erika, on the *Magic*," he said. "She worked in Guest Services. We met in 2007 and married in 2009. We have twins, one boy, one girl, Fabijan and Iva."

(Clockwise, from top left) A stateroom, assembled offsite, is loaded onto the ship; the Hero Zone takes shape; one of the massive LNG tanks waits to be installed; snowfall on the deck (not something that will likely happen in the future); and, for all those hidden Mickey buffs out there, a tangle of cables somewhere on the ship (good luck finding it).

The "heart of the ship," both literally and figuratively, the Grand Hall features not only a stage (a first for a Disney cruise ship), but a balcony for character appearances.

A Hall So Grand
That's Exactly What We Named It

From the magnificent chandelier and the sweeping staircase to the all-new stage and character balcony (and don't forget the bronze figure—actually, figures—at the center of it all), the central gathering place on the *Disney Wish* is so expansive, majestic, and awe-inspiring (not to mention multifunctional) that Atrium—the name used on the existing Disney ships—just didn't quite seem to capture its impressiveness.

"It's the heart of the ship," said Laura Cabo. "It sets the theme for everything guests will enjoy over the course of their journey."

Consequently, Imagineers wanted to bestow it with a new moniker that befits how splendid, elegant, and grand—Wait, that's it: Grand Hall.

Inspired by the enchantment found in Disney fairy tales in general (and *Cinderella* in particular), the space feels like the inside of a storybook castle. Which is no accident.

"It sprang from the idea of a castle," said Cabo. "A castle is the centerpiece of our parks, so it was really kind of obvious to us that we should give Disney Cruise Line its own castle on the seas."

"It is light and airy," she pointed out, "with soaring columns that rise up three stories in the air. The ceiling is covered with decorative ribs. There is a chandelier that sits in the middle of the space and is encrusted with thousands of beautifully sparkling crystals, and at the bottom of that chandelier is a star that signifies the *Wish*."

Of course, this being Disney, even the chandelier has a backstory.

"The shape of the chandelier was inspired by the film *Cinderella*," reiterated Claire Weiss. "Fairy Godmother draws this beautiful wand spiral of magic around Cinderella and her dress becomes real. We took that shape and we made that into the chandelier. Cinderella really is a classic fairy tale about dreaming and wishing, and this is just a continuation of that."

Cinderella herself is centerpiece of the Grand Hall, captured in all the dazzling beauty of her ball gown finery as a bronze figure at the foot of the staircase. Behind her unfolds a little story, also told in bronze, of the mice, Gus and Jaq, seemingly under the protection of Cinderella, but still very much under the wary, watchful gaze of Lucifer the cat, a few feet away under the stairs.

Above and beyond the aesthetic wonders of the Grand Hall are its practical elements, the ones that are felt by guests more than they are seen by them.

"We've really opened up the space," said Mo Landry. "We're not fighting with a queue to get into a restaurant or the elevators. It really gives us the chance to offer those events and activities that we've always wanted to do, and we don't have to worry about, 'Wait a minute, we need this space for people waiting to get into a dining room for dinner.'"

In fact, it's the absence of those aforementioned elevators, which have been relocated to the fore and aft parts of the ship, that have provided the Grand Hall with its most dramatic change from the Atriums on the existing Disney ships.

"The Grand Hall Stage is a game-changer for us," said Denise Case, entertainment creative director, Disney Cruise Line. "We're really able to take our entertainment to the next level because of the stage and the character balcony above it. So, while we do princess gatherings on all the ships, on the *Wish* it goes up a notch because of the [new] venue."

Another way it goes up a notch is when guests board the *Disney Wish* for the first time at the start of their cruise.

"You come in, you're announced, and, new to this ship are our entertainment hosts—we call them our 'shipmosphere' actors—who are there dressed in these really fun overlays that feel very storybook, very fantasy.

"Then, we all say the magic words together, 'Wishes do come true,' and the Grand Hall *reacts*," Case emphasized. "We call this 'welcome aboard' moment 'Once Upon a Wish,' and it's all set to a new song we've written, 'Set Sail.'"

"That first moment when you walk into the Grand Hall is going to be eye-opening one hundred percent," said Landry.

She then pointed to another significant change in the design of the space from the existing ships. "We are also making a statement [with this]," she added. "That we are putting our kids front and center.

"Some people on the team thought we were crazy," Landry recounted. "But then they would think about it, and it would make total sense because we are

The ever-helpful mice friends Gus and Jaq are partially obscured under Cinderella's dress as they search for the glass slipper, but they also need to be on the lookout for Lucifer.

The design process for the Grand Hall involved studying dozens of wildly different color schemes such as this one (top), as well as a number of options for railing patterns (bottom left) and the original inspirational sketch for the three-deck-high columns throughout the space (bottom right).

The Grand Hall was inspired by Disney fairy tales and castles, with the carpet incorporating patterns and icons primarily drawn from *Cinderella* (center left). It's one of the largest interior spaces on a Disney cruise ship (top), with much assembly required, including the sweeping staircase (center right). Imagineers are always checking to make sure reality matches the design (bottom left) while capturing an image of the space before more decks are stacked on top of it (bottom right).

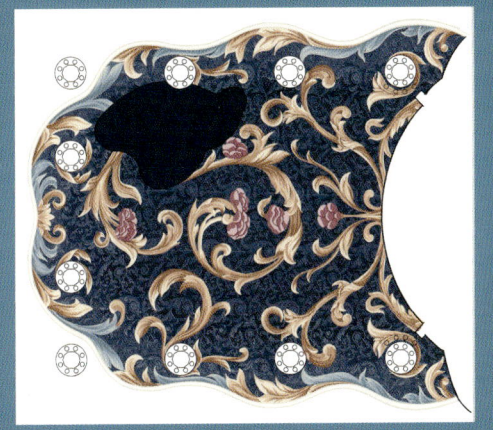

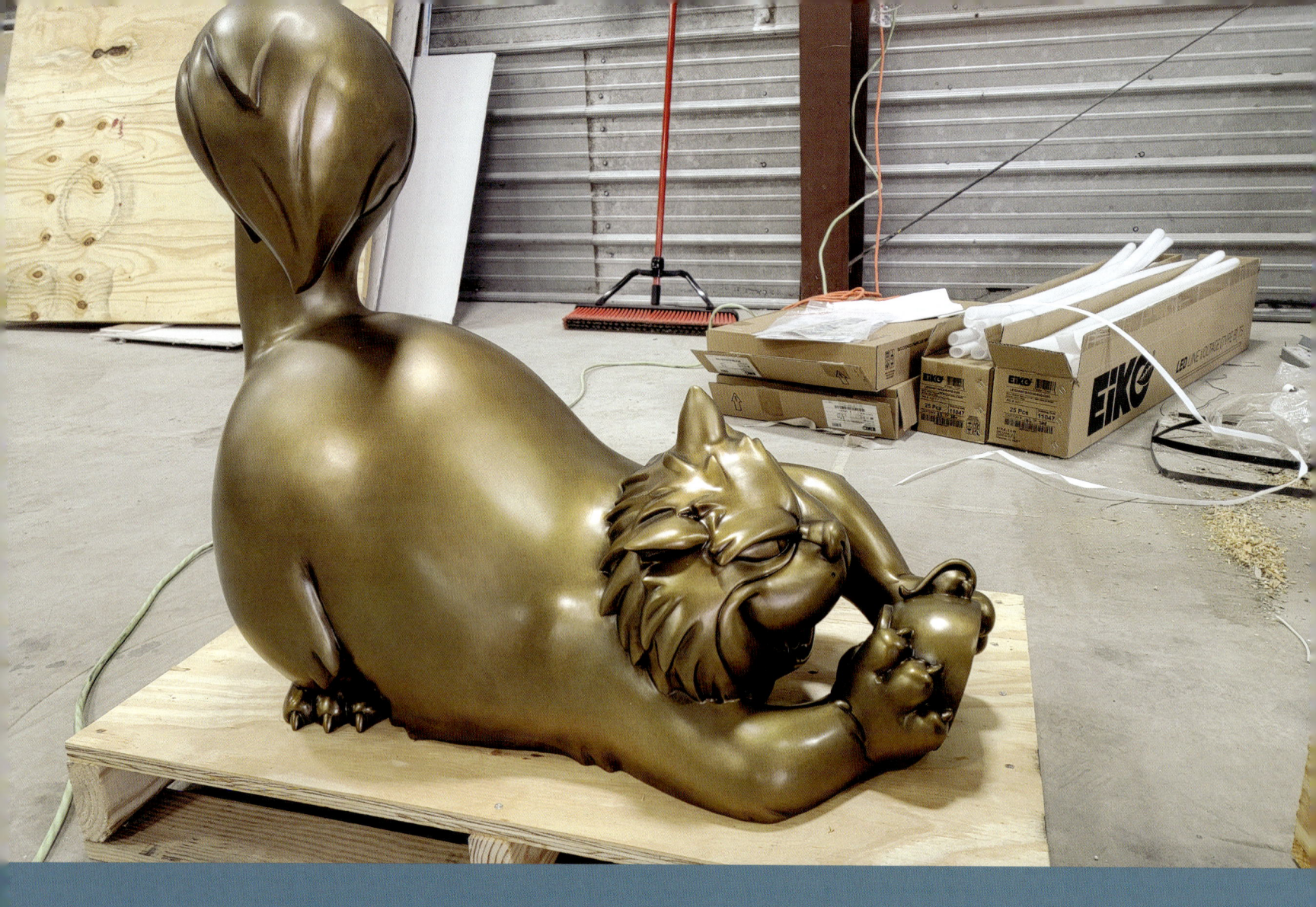

Sculpted and cast in California, Lucifer is ready to be crated and shipped off to Germany, where he will no doubt relish his role as antagonizer of Gus and Jaq aboard the *Disney Wish*.

truly a cruise line for families, and we want everyone to have the best time."

One example of this youth-oriented approach to the Grand Hall, Landry cited, is the check-in to Disney's Oceaneer Club.

"We often joked about the fact that, 'Wouldn't it be nice to just have a slide that the kids could use to come into the youth activities area from other spaces?'" she said. "When we decided to put the Oceaneer Club on Deck 2, it was just perfect to have a slide from the Grand Hall down to the space. We know every parent's going to want to slide down as well and we're going to let them. Tim Hall wanted to know if he could slide down and I said, 'Sure, but I have to go first.'"

The arrangement of Cinderella, Gus and Jaq, Lucifer, and the Glass Slipper in the Grand Hall sets up a familiar story in which the mice would like to find the slipper for Cinderella while Lucifer would like to nab them.

Claire Weiss

Interior Designer Principal, Walt Disney Imagineering

Growing up in Logan, Utah, the closest Claire Weiss ever got to a cruise ship was a point-and-click PC game called *Titanic: Adventure Out of Time*.

"I've always had this weird fascination with ships," she said. "It started as a child—this is before the movie—with the *Titanic*. I went down this long rabbit hole of trying to understand the Titanic from a design perspective. As a twelve-year-old, I read all the documentation that was available, not about the disaster, but really more of the design of the ship."

That fascination fed an interest she already had in architecture, which led her to undergraduate and graduate degrees at the University of Idaho in Moscow ("It was so far north, we were almost to Canada") and undergraduate and graduate degrees in the discipline.

For her master's thesis, she chose to do a redesign of Mickey's Toontown at Disneyland Park.

"I remember there was this point where local architects and all these professors from other schools came in and they critiqued it," she said. "Mine was not well-received. I don't think it was perceived as real design."

While finishing that degree—her thesis was ultimately accepted, by the way—she took a trip to Walt Disney World.

"I was walking through EPCOT and thinking, 'This is incredible. This level of detail is so beautiful.' And something in my brain went, 'That's a job. That could be your job.'"

And it was—six years later.

"I got an email from an old colleague that Disney was hiring to build a park in Shanghai," Claire said. "After being contacted by a recruiter, I was told that Disney didn't have any jobs in architecture, but there were positions in interiors. Would I be interested? I said, 'Totally, but, full disclosure, I've never studied interior design.' That's how I got my foot in the door with Disney."

After working on Enchanted Storybook Castle at Shanghai Disneyland and, with Danny Handke, Guardians of the Galaxy—Mission: BREAKOUT! at Disney California Adventure, she got a call about working on something entirely different.

"If someone had told me as a child that I would be part of this incredible team creating Disney cruise ships for the whole world, I would 100% have not believed them," she said. "My whole career has been a delightful accident, where everything that I intended to do is not the thing I ended up doing.

"You really don't know what's going to happen in life and I'm an excellent reminder of that—in a good way."

Danny and Sachi Handke

Senior Creative Director and Project Coordinator, Walt Disney Imagineering

Once upon a bad presentation . . .

Not exactly the classic opening to a fairy tale love story, but for Danny and Sachi Handke, it will do.

Sachi, having just been named a Disney Ambassador at the Disneyland Resort, was at Walt Disney Imagineering headquarters in Glendale, California, meeting with Imagineers and learning about its upcoming projects.

One of the Imagineers she met was an associate creative director named Danny, who, according to Sachi, gave one of the "weaker presentations" that day.

"In my defense," said Danny, "I was filling in for someone. I didn't really know what I was presenting."

Despite that first impression, the two eventually ended up exchanging contact information. Soon, they were texting, then dating, and in August 2013, less than a year after that fateful presentation, Danny dropped to a knee in the Royal Hall at Fantasy Faire at Disneyland Park and, with all the honor, valor, and nobility of Prince Charming, proposed to Sachi.

Apparently, his presentation skills had improved because Sachi accepted and on March 21 of the following year they married and lived happily ever . . .

Well, the story isn't quite over yet. In fact, there are still many stories to tell.

"I approached this project with the idea of bringing more Disney storytelling into the spaces," said Danny. And that's exactly what he and his team have done. But not just Disney stories.

"We got to work with Pixar Animation Studios, Marvel Studios, and Lucasfilm. It was just so much fun working with all the teams and these great filmmakers."

He also got to fulfill a lifelong dream. "I always wanted to do a Mickey Mouse cartoon," said Danny, who has a bachelor's degree in Media Arts and Animation, "To be able to do these new Mickey shorts in AquaMouse with all the classic characters is a really cool opportunity."

Sachi is also on the project, part of the team focused on creating and completing the media that goes into AquaMouse, Worlds of Marvel, and *Star Wars: Hyperspace Lounge*.

"This project is all hands-on deck, pardon the pun," she said. "Everyone has each other's back and it's so cool to work with people who are really good friends."

And, in some case, spouses, who will, we are told, live happily ever after.

With design inspired by the song "Oh Sing Sweet Nightingale" from *Cinderella*, Nightingale's is a piano bar off the Grand Hall that by day is part of the Grand Hall, but at night becomes an intimate, more adult-focused environment.

What to Expect When You're Exploring

If you've been on one of the existing Disney cruise ships, don't be surprised if it takes you a little time to get the lay of the land on the *Disney Wish*. It's not that its layout is confusing—far from it! It's just that this is not your previous Disney cruise ship. "The *Disney Wish* is truly a whole new spin on our ships, not only in what we are offering from an experience and design point of view," said Laura Cabo, "but also from the fundamental layout of the ship."

Some features and facilities were moved to more ideal places, while others were completely reconfigured. And then there are the entirely new spaces, created to provide fresh, original experiences for Disney Cruise Line guests.

"By really integrating the spaces more, we've improved guest flow throughout the ship," said Mo Landry, director, Entertainment Operations and Special Projects, Disney Cruise Line. "On the existing ships, we've had parents who, after dropping off their kids at youth activities, don't know what to do. Now everything is right up the stairs. Nightingale's . . . it's ready for you. At Keg & Compass, the big game may be on.

The clubs and lounges on the *Disney Wish* all draw on Disney stories for their inspiration. "The way we impart the story for these spaces is not through the story beats, but through feelings and emotions," said Claire Weiss. "When I watch those films, what are the feelings I have? What is my takeaway?"

Take The Bayou, a centrally located lounge on Deck 3 that draws its design cues from *The Princess and the Frog*.

"Yes, it's about Tiana," said Weiss. "Yes, it's about her dream. But it's also

Deck Ten 97

a story about light, a story about color, a story about the natural environment. You take those elements and create an environment that makes you feel the way the movie made you feel."

The result is a lounge evocative of the magical swamp from *The Princess and the Frog*. Overhead is a canopy that appears to be magnolia blossoms and Spanish moss, aglow with the twinkling of dancing fireflies. Lily pads are a recurring motif, seen in light fixtures, the carpet, and even a bronze sculpture of Tiana and Naveen in their frog forms.

Nightingale's is a piano bar on Deck 3 just off the Grand Hall that pays homage to the scene in *Cinderella* in which the two stepsisters, Drizella and Anastasia, and their mother, Lady Tremaine, attempt to perform "Oh Sing Sweet Nightingale." The space is decidedly more lovely than the stepsisters' rendition of the song: classically modern with hints of colors and patterns from the film, which makes it more in line with the beauty and elegance of Cinderella's version of the song that ends that scene in the film.

Also on Deck 3 is Keg & Compass, a pub-like space that celebrates the romance, adventure, and stories of the seas in a manner that is both imaginative and imagined. Architecturally, the space takes inspiration from a sailor's map room, with walls and columns in the slightly rusticated style of late 1800s Norwegian shipbuilding. The furniture and bar counter are designed to resemble map storage drawers.

Of particular note is the *Seafarer's Map of the World*, a massive atlas that stretches across the entire ceiling and is illustrated with sea-related and sea-adjacent characters, as well as references to Disney history, place-making, and folklore.

Period-style oil paintings on the walls expand upon the nautical tales found on the ceiling, while the porthole frames are accentuated with intricate carvings: octopus tentacles, barnacles, compasses, and other references to the ocean and oceanography. There's even a ship in a bottle (which looks suspiciously like the *Disney Wish*).

Perhaps the most unique bar on the ship is the one that is set in a place from "a long time ago in a galaxy far, far away. . . ."

"The concept simply came from the idea of adults wanting to hang out in the *Star Wars* universe," said Danny Handke.

"We always talked about the fact that kids get to immerse themselves in *Star Wars*," he observed. "But as a grown-up, and a parent, I'm jealous that I don't have my own space on the ship where I can do the same. So we pitched this idea: 'What if we did a lounge that's an adult playground for the world of *Star Wars*.' And everyone just loved that idea."

Far from the seediness of planet-bound bars such as Oga's Cantina at *Star Wars*: Galaxy's Edge, *Star Wars*: Hyperspace Lounge is styled as the interior of a yacht-class ship on par with *First Light*, Dryden Vos's ship in *Solo: A Star Wars Story*. It features a lavishly appointed interior of polished metals, rich leather, gold furnishings, and fine art pieces and props.

Taking its design cues from *The Princess and the Frog*, The Bayou is an informal retreat in the midst of the shop-lined corridor leading from the Grand Hall to the Walt Disney Theatre.

Perhaps the most anticipated new lounge on the *Disney Wish* is *Star Wars*: Hyperspace Lounge, taking guests on a tour of the galaxy aboard a yacht-class ship. The space was completely built offboard (top) before being packed up and shipped to the ship. Wanting everything right, designers provided extensive notes on renderings (bottom left). Laura Cabo (left) and Nick Snyder check the lounge's window on the world of *Star Wars*.

"We've created four 'hero' props for the lounge," said Kristen Zeigler, set decorator, Walt Disney Imagineering. "They're showcased in these dome displays that are similar to what you would see in Vos's study on board his yacht."

The pieces are a hallikset, a musical instrument, similar to a guitar, from the planet Naboo; a small statue of a Togruta, a humanoid species from the planet Shili; Mustafar lava crystals; and a stuffed hawk-bat. "It looks like this taxidermy creature that we've never really shown in the films or lands in our parks," said Zeigler.

A large "window" over the back of the bar looks out into space upon various destinations in the *Star Wars* universe as the "ship" jumps through time and space to Coruscant during the Republic Era, Mustafar of the Empire Era, Tatooine of the Mandalorian Era, and moons of Endor and Batuu of the First Order Era.

During the day, the lounge offers a low-key, family-oriented atmosphere, but after dark it turns into an adults-only venue with more of a club vibe that includes enhanced lighting effects and a more energetic and idiosyncratic cantina-flavored soundtrack, not to mention craft beers and cocktails "locally sourced" to each of the places the ship visits.

Those familiar with the D Lounge on the existing Disney cruise ships should recognize the Triton Lounge on Deck 5. "It's great for smaller, more intimate experiences, such as presentations, premium tastings, and small functions like birthday or engagement parties," said Denise Case. "The learn-to-draw program is also here. We've added a video component so that now guests are actually going to be taught how to draw Disney characters by a Disney artist. It feels like the artist is live and it's happening in real time."

Change has also come to the cinema—or should we say cinemas?

The *Disney Wish* features two boutique screening rooms on Deck 4 that not only offer a more intimate, immersive moviegoing experience, but more film choices as well (since that's the only thing that will be in these spaces).

The décor of the portside Wonderland Cinema is inspired by the fantastical world of *Alice in Wonderland*, while the Never Land Cinema, on the starboard side, takes guests on their own flight of *Peter Pan*-informed fantasy, past the Second Star on the Right, and straight into the magical world of movies.

It's the new-to-the-Disney Cruise Line spaces, though, that provide the most dramatic impact in terms of layout, scale, and entertainment programming.

If the Grand Hall is the "castle" of *Disney Wish*, Luna is the "town square," an event-gathering place on Decks 4 and 5 that is architecturally inspired by London's Globe Theatre and named for a 2012 Disney•Pixar short.

"Luna really opens up, with that double-height space, the ability to do popular games and activities with the entire family in there," said Landry. "Then later at night, we turn it into the adult game show area and a nightclub to really give those groups the opportunity to have fun."

The multipurpose space has the ability to host everything from bingo and trivia contests to performers such as magicians and comedians.

Bob Girardi thought of one act in particular that should see some benefit from being able to perform in a venue with such a high ceiling. "Now that we have a two-story space, I'm sure the jugglers are going to like that," he said.

"Luna is this amazing venue that is going to make our events and experiences even better," said Denise Case, "because we now have a video screen and a video wall and other technologies that enable us to do things we've never been able to do before."

"Let's take karaoke, for example," she continued. "When guests get up there to sing, that video wall is like *American Idol*. It gives you a backdrop. So if you're singing 'Can You Feel the Love Tonight,' [from *The Lion King*], there'll be this beautiful starry sky behind you.

"You're not just in a club anymore," Case pointed out. "You're literally in an environment because it's all about enchantment and magic coming to life and being immersed in the story."

Game shows continue to be a big part of the entertainment option aboard Disney's newest cruise ship as well. "One of our new shows is *Villain Game Night*," said Case. "It's the Disney villains kind-of, sort-of playing a version of a *Hollywood Squares*-type game show."

And of course, there's still room for an old favorite. "Match Your Mate is here," she said, "but in the spirit of diversity and inclusion we made some changes. We've taken out all the husband-and-wife references and now it's your partner," she continued. "Everything about that show is not whether you are married to a man or a woman. It's about the couple and the partnership and the time they've been together. It's important that all of our guests feel like they belong."

The other new venue, on Deck 14, is also a two-deck multipurpose space, but the activities here are usually going to be a little more physical in nature.

"Everyone has compared Hero Zone to a sports deck," said Landry," but it's so much more. That's why we refer to it as an entertainment activity zone versus a basketball court, not that anything is wrong with basketball. We want our guests to be active, and they can still do that. But it's not just about basketball. It's being able to engage every guest."

Landry also pointed out that this is an indoor space, noting the advantages that affords those who use it.

"The fact that it's air-conditioned is amazing," she said, "as is the fact that there's no wind [sending] Ping-Pong balls or basketballs into the ocean. That's a huge plus."

Senses Fitness provides a comfortable backdrop for those wanting to keep fit while onboard (top). After a vigorous workout, guests can take in a first-run film at one of two intimate, immersive screenings rooms: the down-the-rabbit-hole-vibe of Wonderland Cinema (bottom left) or the London-to-Lost-Boys-ambiance of Never Land Cinema (bottom right).

An obstacle course themed to *The Incredibles* is one of the activities on tap in the new Hero Zone, a two-deck, climate-controlled multipurpose space.

Denise Case describes one of the areas this way: "a giant inflatable obstacle course," she said. "Think *Wipeout* themed to *The Incredibles*. Guests can sign up and compete, or they can just watch from the balcony."

"Also," Case added, "because it feels like a gymnasium, we can easily have dance parties for teens and tweens that feel like school dances."

But probably the most popular event?

"On our existing ships, we do Jack-Jack's Incredible Diaper Dash in the Atrium," Case said. "Here, we've moved it to the Hero Zone. We can accommodate more spectators and way more babies."

One last thing that won't be found in its old familiar place is Senses Spa. The immensely popular rejuvenation and revitalization spot, an upper deck mainstay on the four existing Disney ships, is on Deck 5 of the *Disney Wish*, and features two relaxation spaces. One is a rain forest–themed indoor area with an ice room, an aromatherapy steam room, a dry sauna, and a calming pool. The other relaxation space is actually outdoors, in the ship's bow, with whirlpools, lounge chairs, and plenty of sun, although there are also tented structures and nooks to provide shade.

Deck 5 also offers other "self-improvement" diversions at Senses Fitness, featuring exercise equipment, weight lifting machines, and free weights, as well as a dedicated cycling studio, plus exercise and wellness rooms.

To top it all off, there are separate salons for adults and children.

Untangled Salon takes its inspiration from a Disney character well known for her beautifully flowing—and very long—trusses. Just take a look at the back of the ship, where she appears with her pal, Pascal, painting the ship's name and filigree.

Like Rapunzel, guests can let their hair down here—and have it styled any way they want. (They can also opt for manicures, pedicures, and skin treatments.)

Hook's Barbery brings to mind a traditional European men's salon, this one in the spirit of that man among men, Captain Hook, and his personal quarters aboard his brig, the *Jolly Roger*. In addition to hairstyling, shaves, and nail and skin care, the Barbery boasts the ultimate toast to a pirate's life: a hidden bar, concealed by a heavy wooden door on which Captain Hook's private map of Never Land is displayed.

And, just as the Fairy Godmother turned Cinderella into a beautiful belle of the ball, girls and boys can enjoy a magical makeover at the Bibbidi Bobbidi Boutique. This regal salon enables kids to live their storybook fantasies, whether it's as a fairy-tale princess, a valiant knight, a swashbuckling buccaneer, or perhaps even a youngling or Jedi master.

Because once you're made up however or as whatever you want to be, there are plenty of opportunities for a night out on the *Disney Wish*.

The *Disney Wish* offers salon services for every member of the family, from the warm-gray-and-dark-wood tones of Captain Hook's Barbery (left), to the light, airy atmosphere of Untangled Salon (below left), to the regal, fairytale-inspired Bibbidi Bobbidi Boutique (right).

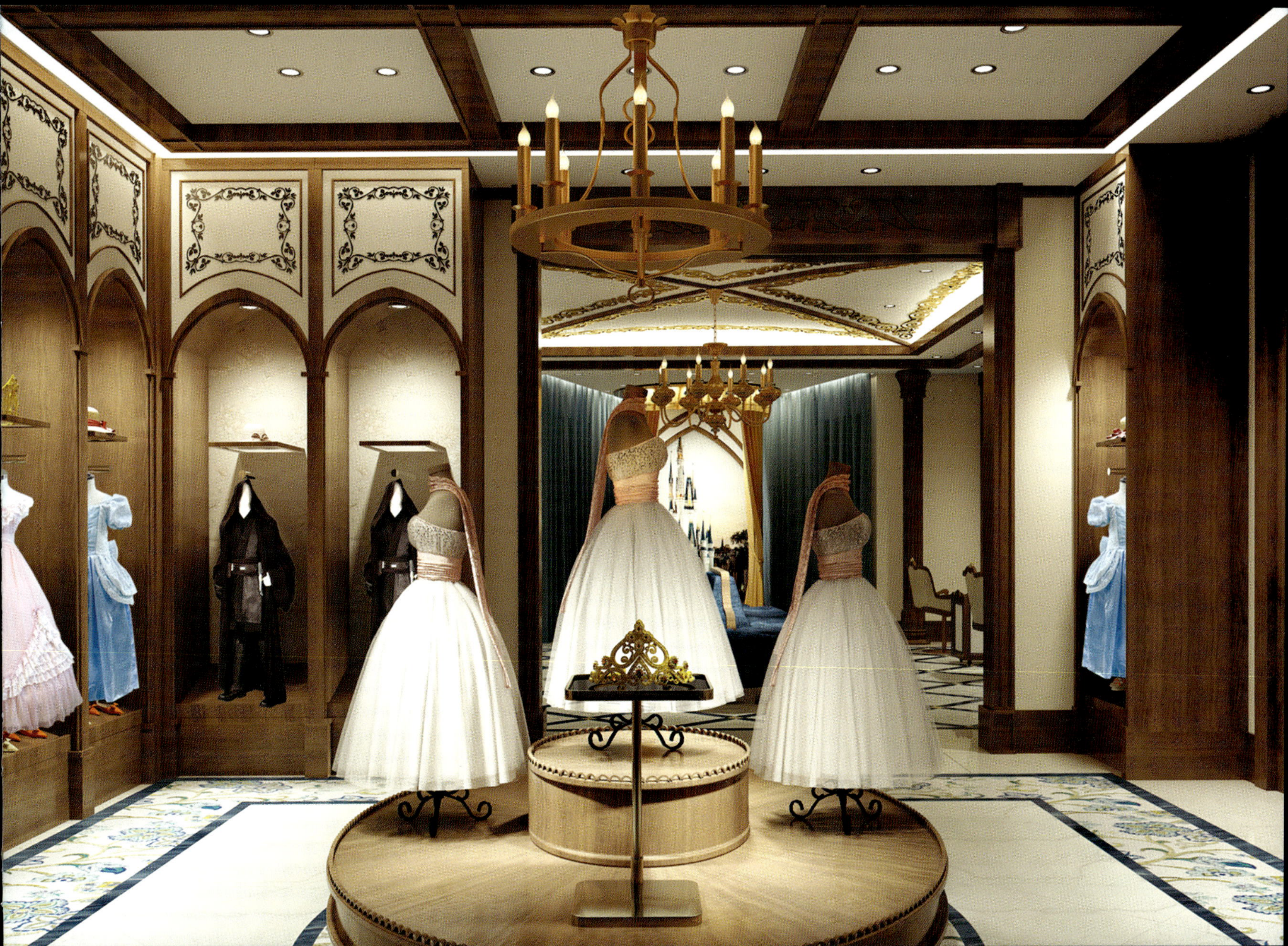

Ashley Long Cruise Director, Disney Cruise Line

For Ashley Long, performing was always her first love.

"I spent much of my childhood and adolescence onstage dancing, acting, and singing," she said.

Disney was also a big part of her life. "Our first family vacation was to Walt Disney World," she said. "My sisters and I ran from one ride to another, trying to squeeze in as much as possible."

It seemed inevitable, then, she would one day end up at Disney. All it took was the summer after she graduated from college performing for cruise ship guests visiting Skagway, Alaska, to convince her that she should actually be working on a cruise ship. She applied for a position with Disney Cruise Line and was hired.

More than a decade later, she has served as cruise staff, entertainment and activities manager, youth activities manager, assistant cruise director, and, today, cruise director on the *Disney Wish*.

"I can't imagine doing anything else," said the Bowling Green, Kentucky, native, who, when she's not on a Disney ship, still lives a mile from her parents.

Jimmy Lynett Cruise Director, Disney Cruise Line

Disney was a big part of Jimmy Lynett's life growing up. "My parents took my siblings and me to Walt Disney World," he said. "We went pretty regularly and loved it."

It should come as no surprise, then, that Jimmy wanted to one day work for, as he calls it, "the mouse." What may be surprising though, is that he also wanted to work on a cruise ship; this from a kid who grew up in Scranton, Pennsylvania.

In 1998, Jimmy's worlds collided with the impending launch of the *Disney Magic*, the first ship of the new Disney Cruise Line. "I applied many times," he said, "and was finally successful in December 1998. I've been a very proud crew member ever since."

Over the years, Jimmy has served as a youth activities counselor, port adventures manager, assistant cruise director, cruise director, and, today, *Disney Wish*'s entertainment operations leader.

He also met his wife, Zoe, while the two were working on the *Disney Magic*. They and their three children (Jamie, Mia, Drew) live on the Gold Coast of Australia, where his wife grew up.

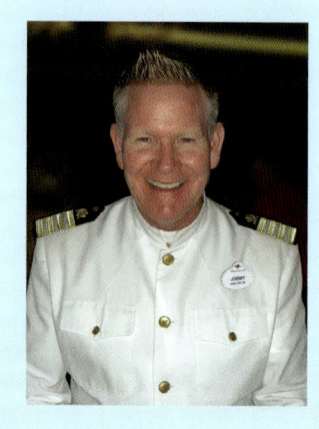

The nautically inclined Triton Lounge is a presentation space for shopping previews, cooking classes, alcohol tastings and pairings, and a variety of other demonstrations and events.

Picking up the story of Anna and Elsa after the events of *Frozen II*, Arendelle: A Frozen Dining Adventure takes place in the kingdom's magnificent hall. It's an engagement party for Queen Anna and Kristoff and everyone is here to celebrate with a splendid meal ("catered" by Oaken) and a bevy of special guests, including Olaf and a surprise appearance by . . . well, she's pictured above.

Eat, Drink, and See Elsa...

...and Anna and Ant-Man and the Wasp and Mickey and Minnie and the classic Disney Studio in Burbank. Dining on a Disney cruise ship is an adventure in and of itself—and the *Disney Wish* is no exception. The rotational restaurants (an innovative concept that started with the *Disney Magic* and remains exclusive to the Disney fleet) boast displays of entertainment, elegance, and technology, but that's just the start of the gastronomic options that range from the casually whimsical to the whimsically elegant.

"What I love about the three rotational restaurants on the *Disney Wish* is that each one has such a different flavor," said Danny Handke. "Every night of your cruise, you're going to get such a different experience. Arendelle is a live entertainment experience, Worlds of Marvel is a technology-level experience, and 1923 is an elevated, elegant experience; so there's going to be something for everybody."

Let's start in 1923 (the restaurant, not the year): "1923 is a celebration of the Disney Studio," said Claire Weiss, "[that's] focused on artwork and animation. But we want you to feel as you're having that vintage Hollywood feeling that people who have lived in Los Angeles know and understand just by going to Musso & Frank or popping into the Frolic Room." (The name, by the way, refers to the year The Walt Disney Company was founded.)

Guests dine in one of two areas: the Walt Disney room or the Roy Disney room. Both offer interior settings of warm woods and elegant metal grillwork that evokes the Golden Age of Hollywood glamour. Amid rough drawings, storyboards, backgrounds, cels, maquettes, and props that celebrate a century of Disney animation, guests are invited into the process, learning how Disney artists take their ideas and stories from sketch to screen. The focus, of course, is on enchantment (the theme of the *Disney Wish*) and among the many animated classics featured are *Snow White and the Seven Dwarfs*, *Cinderella*, *Sleeping Beauty*, *The Little Mermaid*, *Beauty and the Beast*, and *Frozen*.

The fact that there are two discreetly separate dining rooms is a new twist on the rotational restaurants. "We thought maybe the restaurants were *too* large," said Laura Cabo, "so we broke up two of them into two separate dining rooms to create more intimate dining experiences. Now you dine with 350 other guests, rather than seven-hundred.

"The exception to that is Arendelle," she provided, "where we wanted everyone together in an energetic theatrical space."

Styled as a grand dining hall in the castle in the kingdom of Arendelle, the restaurant picks up the story of Anna and Elsa after the events of *Frozen II*. Queen Anna and Kristoff are having an engagement party, "catered" by Oaken's Hearty Party Planning Services.

"We said, 'Well, it was so wonderful how Kristoff was trying to propose to Anna, but let's not do the wedding, let's do the big engagement party instead,'" said Ed Whitlow, director, artistic and talent casting, Disney Cruise Line. "And that's what it's become."

Whitlow has experience staging this type of "dinner theater." He wrote and directed the show for Tiana's Place on the *Disney Wonder* (and also, "in another lifetime," he said, performed in the Hoop-Dee-Doo Musical Revue at Disney's Wilderness Resort at the Walt Disney World Resort).

"Tiana's Place was really our first foray into doing entertainment restaurants on the ships," said Whitlow. "We learned how a show like this fits into a venue that's really all about family and dining. We wanted to support it with entertainment, not overwhelm it."

The same is true for the celebration in Arendelle, but that doesn't mean it won't have its share of showstopping, sing-a-long moments—along with one particularly magical innovation.

"The big takeaway on this one is the technology that has been created for Olaf," said Denise Case. "That means he'll be interactive and to scale, not this big walk-around."

Speaking of technology, the Worlds of Marvel restaurant is all about pulse reactors, high-definition 3D displays, and quantum cores, which shouldn't seem too surprising when people such as Tony Stark, Hank Pym, and Carol Danvers are involved (allegedly).

"I've always wanted to be a part of a Marvel-level action show like this," said Danny Handke. "One that stars the Avengers and blows guests away with its enormous production values, massive special effects, and live elements."

Plus, you get dinner.

Named for the year The Walt Disney Company was founded, 1923 embodies the spirit and atmosphere of the Walt Disney Studios in an elegant, sophisticated setting amid drawings, storyboards, backgrounds, cels, maquettes, and propping that celebrate a century of Disney animation.

Worlds of Marvel offers both dinner and a demonstration of quantum technology by Ant-Man and Wasp (top left). The presentation has been dubbed "Avengers: Quantum Encounter" (right) and, when the exhibition goes slightly awry (center left), our heroes are here to save the day—with a little help from guests.

The entrance (bottom left) is sleek and austere, punctuated by Art Deco-style bas reliefs of the Avengers.

Every table in the restaurant comes with a quantum core, pulsing with energy and equipped with interactive buttons that will be used during the demonstration (bottom right).

Worlds of Marvel is a state-of-the-art mobile base of operations for the Avengers that serves as both a technology showcase and a table-service restaurant.

As the story goes, guests are not only here for a meal, but also to participate in Avengers: Quantum Encounter, a demonstration of quantum technology being staged by Ant-Man and Wasp. When the exhibition goes slightly awry (as these things inevitably do, especially when Ultron has infiltrated the network), Ant-Man and Wasp end up calling on Captain America, Ms. Marvel, Captain Marvel, and even the diners themselves to help save the day.

Guests don't always have to come to the world's rescue to eat well aboard the *Disney Wish*, though. The ship features a *bounty* of other options, ranging from a quick meal to a luxurious dining experience crafted by a 3-Michelin-starred chef.

The former can be had at Mickey & Friends Festival of Foods on Deck 11, where five walk-up food stalls offer classic fare, everything from cuts of barbecue beef, pork, and chicken to pizza, hamburgers, hot dogs, tacos, and soft-serve ice cream.

Also on Deck 11 is Marceline Market, an indoor food court (with seating both inside and out) that's named for Walt Disney's early childhood hometown in Missouri. The space is styled as an old industrial loft with proprietors straight out of Disney stories. A map at the entrance serves as a guide to the seating areas, which include a farmer's market featuring Wicked Witch Produce, the Hopps Family Farm, and Flower's Flowers (*Bambi*); bakery and cooking supplies with Atilla the Bun; and toys and inventions created by Geppetto and Maurice.

Remy provides the motto for the market, "I ONLY WANT TO EAT THE GOOD STUFF," captured in a neon sign, also near the entrance.

The really, really good stuff is on Deck 12. That's where guests can find a pair of premier dining establishments exclusively for adults. One is centered around "the Casanova of candelabras," while the other sort of pays homage to a majordomo-turned-enchanted clock.

"In both of these restaurants, which are for adults, you're not going to see *Beauty and the Beast*, the animated film," said Pam Rawlins. "What you are going to see are designs that have been inspired by the film that are gorgeous and refined. It invokes a feeling, and I believe our guests are going to be extremely surprised when they see these spaces.

"That's the tone we've set," she added. "That you can do a very elegant restaurant like Enchanté or a steak house like PALO and be inspired by the animation of *Beauty and the Beast*; sophisticated and stylized."

Enchanté features a French-inspired menu crafted by 3-Michelin-starred chef Arnaud Lallement. It's also under the watchful eye of maître d' extraordinaire Lumiere, whose introductory greeting to Belle serves as the name of the restaurant.

PALO Steakhouse is a new twist on a Disney Cruise Line favorite, with subtle nods to Cogsworth and his sense of propriety, properness, and Britishness, not to mention his obsession with clocks.

Speaking of time, the shows that can be found aboard the *Disney Wish* are about to start.

The most luxurious dining experience on board, Enchanté (opposite page) features a French-inspired menu crafted by 3-Michelin-starred Chef Arnaud Lallement.

Marceline Market is a food court styled as an old industrial loft (the name is a reference to Walt Disney's early childhood home in Missouri). The dining areas offer stunning ocean views (top left). The final version of the Marceline Market directory will charm guests (center left).

For a quick bite to eat on the upper decks, there's Mickey & Friends Festival of Foods, featuring a cavalcade of Disney character walk-up counters (bottom) serving everything from hamburgers and hot dogs to pizza and tacos.

Another inadvertent hidden Mickey (far right), this one found by Claire Weiss (those are her boots) is in the floor of Marceline Market, but is not visible to guests (other than in this exclusive photo).

Stephen Walker Culinary Director, Disney Cruise Line

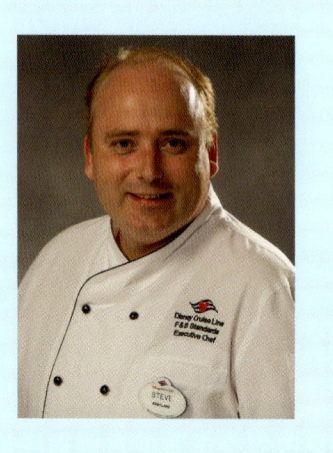

Steve Walker was working as a senior sous chef in London when he saw a newspaper ad for a "New Disney Cruise Ship."

"I was always interested in working on ships," he said. "The fact that it was Disney piqued my interest. It was an opportunity too good to pass by."

Especially for someone who always felt he had an affinity for Disney, even as a young boy growing up in Blackwaterfoot on the Isle of Arran, Scotland. "I was captivated watching *The Wonderful World of Disney*," he said.

In 2000, Steve joined Disney Cruise Line and soon became the executive chef on the *Disney Magic* and *Disney Wonder*, where he worked for seven years before moving on to serve as the head chef for other cruise lines. He returned in 2009 to help launch the *Disney Dream* and *Disney Fantasy*. "I just couldn't pass up another opportunity to work for Disney Cruise Line again," he said.

Today, the Winter Garden, Florida, resident oversees the culinary program for Disney Cruise Line including, menu development and implementation, chef training, and consistency maintenance.

Marco Nogara Captain, Disney Cruise Line

The *Disney Wish* may be brand-new, but it will be in the very capable hands of a veteran captain with more than four decades of experience at sea.

Born and raised in Venice, Italy, Marco Nogara was educated at the Nautical School of Venice and the Italian Naval Academy in Livorno.

After serving as navigating officer on a destroyer in the Italian navy and working on several cruise lines, it seemed inevitable that he would eventually join Disney Cruise Line.

"The very first movie I ever saw was *Mary Poppins*," said Marco. "And, like most kids in Italy, I grew up reading *Topolino* [*Mickey Mouse*] magazine."

In 2010, he was hired as a captain on the Disney Cruise Line fleet, where he has taken charge of all the Disney ships: the *Disney Wonder*, *Disney Magic*, *Disney Dream*, *Disney Fantasy*, and now, the *Disney Wish*. When he's not aboard a Disney ship, Marco lives in Venice with his wife, Pamela; son Nicolo, and two French bulldogs, Morgana and Decimo.

The name may sound familiar to Disney Cruise Line veterans, but the experience is much different at this PALO. It's a steakhouse that owes much of its inspiration to Cogsworth, the British-seeming majordomo-turned-enchanted clock from *Beauty and the Beast*.

Home to Broadway-caliber productions and motion picture prem "ears," the Walt Disney Theatre finds its design inspiration in the fantastical forest worlds of the Disney animated classic *Fantasia*, especially in "The Nutcracker Suite" and the "Pastoral Symphony" sequences.

All the Ship's a Stage
(But Mostly the Walt Disney Theatre)

"We are always trying to do something that is truly Broadway caliber," said Ed Whitlow when commenting on the shows in the Walt Disney Theatre aboard the Disney Cruise Line ships. "They are the real thing. Any Broadway house or West End house would love to have a show of our quality." The Walt Disney Theatre on the *Disney Wish* is no exception, with two entirely new shows and a third that has been updated and reimagined.

But first, there's the theater itself.

"It was very innovative at the time on those first ships [*Disney Magic* and *Disney Wonder*]," said Bob Tracht. "It's still amazing to me after all these years that we had the most technically advanced theaters at the time, even more so than in our parks. It was the most sophisticated—and it was on a cruise ship.

"As the technology has improved over the years," he continued, "we've been able to take advantage of that and plus it up with every ship we build."

Disney Cruise Line's iconic theater has a slightly different look on the *Disney Wish*. Finding its design inspiration in the fantastical forest worlds of the Disney animated classic *Fantasia*, the theater boasts a palette of rich forest greens, soft sages, warm woods, regal marbles, and gilded columns and trims, overflowing with dimensional flowers and leaves.

While designed to be timeless, the space has been outfitted with more modern features, such as projection mapping, and—in perhaps the most welcome news of all to those who have been on the existing ships—the entrances to the restrooms are actually inside the theater.

It's one thing to design and build the theater; it's another to fill it with the world-class productions that Disney Cruise Line guests have come to expect.

"It was May 15, 2019, when we had our first creative brainstorming," remembered Whitlow. "I had started a little sooner than that, but this was when we pulled together a core team of fifteen people to do some brainstorming around what kind of entertainment we had done and what we wanted to do now."

The next six months were spent coming up with ideas and narrowing in on the concepts that would be developed into the three productions for the *Disney Wish*.

"We had a big cruise directors' summit in the fall of 2019," said Whitlow. "That's when we took a lot of the ideas and presented them. Cruise directors know what's going on with guests. They know what they like and how they're going to respond, so that was very, very helpful, especially when it came to the new 'welcome' show."

Current opening night shows on the existing ships focus on "promoting what the cruise is going to be about, what's going to happen, where we're going, making sure you go to this place or that place on the ship," according to Whitlow.

"This show is all new, and the unique thing about it is that we're telling a true, original story with our characters," he emphasized. "It's about Goofy and how he wants to be a captain. Captain Minnie, who is so important to the *Disney Wish*, mentors him and sends him on a journey to understand that he can be a captain, but there are some things he needs to learn from other Disney stories."

Thus begins a journey through such Disney films as *Frozen*, *Moana*, *Brave*, and *The Princess and the Frog* in which Goofy learns to believe in his dream, culminating in his becoming an honorary captain.

"And then we do something we've never done before," said Whitlow. "To celebrate, we ask the audience to follow the characters out to the Grand Hall for a huge meet and greet. It takes a lot from a logistics standpoint to manage such an event on a first night, but we think it's really special for our guests."

As for the other two nights in the Walt Disney Theatre?

"The one thing our guests love are the 'book' shows of the really popular films: *Aladdin*, *Beauty and the Beast*, *Frozen*, *Tangled*, *Cinderella*," said Whitlow.

The *Disney Wish*, of course, is no exception.

One of its shows is *Aladdin*, which, according to Whitlow, has been "kind of reimagined for the *Disney Wish*. How? "The Walt Disney Theatre is a bit different on the new ship," he noted, "which prompted us to make some significant scenic changes. We've also been influenced by the live-action version of the film. The new designs have a more textural feel to them. They feel less animated, more real and three-dimensional. We've got a huge LED screen upstage that brings

Disney SEAS THE ADVENTURE!

The opening night show in the Walt Disney Theatre on a *Disney Wish* cruise is "Disney Seas the Adventure" (do you see what they did there?), an all-new production in which Goofy, of all characters, dreams of being a captain on the ship. Does Goofy get his "Wish"? Okay, you may not be entirely surprised by the end of his story, but you will be surprised by the end of the show.

Disney THE LITTLE MERMAID

Created exclusively for the *Disney Wish* is a brand-new production, *The Little Mermaid*. The lavishly staged show is a new take on a classic Disney tale, told in a contemporary way with a bold approach to story and costumes.

©Disney

all the backgrounds to life like you wouldn't believe."

The other show is brand-new, created exclusively for the *Disney Wish*.

"*The Little Mermaid* is a great, great story for us to do," said Whitlow. "Our world today is changing how we tell our stories. With this one, we're taking different approaches to the creative and I love that.

"The big difference is in giving Ariel the agency she didn't have before. The story is told through storytellers. It has sort of a *Once on This Island* kind of feel," he said, referring to the Tony Award-winning Broadway musical. "Our show also has a contemporary tone to it, in the costuming, that will be relatable to the kids who grew up with the movie and who have kids of their own today."

The Walt Disney Theatre, though, is not the only stage on the *Disney Wish*. There's entertainment on practically every deck, in restaurants, lounges, clubs, and even on the top decks.

Especially on the top decks.

"On the *Wish*, the top decks are tiered," said Bob Tracht, "which has created more viewing spaces for live performances."

Added Mo Landry, "This new configuration is awesome and the viewing it's created for our nighttime events and our sail-away parties has made it even more amazing, because our guests don't have to struggle to find the right place anymore. There are now so many *right places* for them to be."

What kind of entertainment will guests find on the top decks?

"Some of what we offer on the *Disney Wish* is similar to our other ships," said Denise Case. "The deck parties, like the sail-away experience? They're a staple of what we do."

But the more things stay the same, the more they change. At least on this ship.

"We decided to continue with a pirate deck party because you're in the Caribbean on this ship and it just feels right," said Case. "But we've really switched it up.

"Now it's called Pirate Parley," she explained, "and we've taken it to the next level with the addition of a live rock 'n' roll pirate band, which I think goes together beautifully. Nothing says rock 'n' roll like pirates."

The show stars Captain Redd, and centers on the redheaded pirate who has taken on a more prominent role in the Pirates of the Caribbean attractions in the Disney parks.

"Our story has gone back to the attraction," said Case. "Guests still get the pirate experience that they've always known and loved, but it's a different pirate experience."

For Case and the rest of the Disney Cruise Line entertainment team, the *Disney Wish* offers unique opportunities for telling stories that don't exist anywhere else. "We have our guests with us for several days on the ship," she said, "so we can build some of these really fun experiences where we bridge the storytelling from when they first step on board to when they leave. We can really build the momentum and the flow of the experience over the length of the cruise. That's a nice thing to be able to do."

Performers rehearse a scene from *The Little Mermaid* at the Disney Cruise Line live entertainment facility in Toronto.

The Walt Disney Theatre takes shape on the *Disney Wish*. The theater is even more massive than it appears when finished, with not only seating for 1,274 guests, but a backstage that's the envy of many Broadway and West End stages.

Denise Case

Director, Entertainment Creative, Disney Cruise Line

Denise Case didn't spend a lot of time figuring out what she wanted to do with her life.

"I was seven years old when I saw my first ballet," said Denise, who grew up in Edison, New Jersey. "Right then and there, I knew, 'I'm going to be a dancer.'"

It didn't take long for Disney to enter the picture. Her first professional job, at the tender age of 19, was dancing at the Walt Disney World Resort in the Kids of the Kingdom, an energetic troupe of talented singers and dancers who performed at the Magic Kingdom.

"I was the quintessential musical theater performer" she said. "I sang a little, danced a lot, carried a tune, acted, the whole nine yards."

Her talents took her to such places as New York and Tokyo Disneyland before she married Joe Case (who heads up the Show Management department for Walt Disney Imagineering), settled in Florida, and transitioned into choreography.

She's been part of the opening teams for the *Disney Dream*, creating and staging the deck parties, game shows, and character experiences that are such a big part of a Disney cruise, and Shanghai Disneyland, which included the castle show, "Golden Fairytale Fanfare." She also worked on a number of shows and events around Walt Disney World.

In 2019, she joined the *Disney Wish* team.

"Everything we do has to have that Disney touch to it," she said. "You can only see it and experience it on a Disney cruise ship and that's what makes it so challenging—and so special."

Ed Whitlow Director, Entertainment Creative, Disney Cruise Line

Ed Whitlow has been in the entertainment business, singing, dancing, and acting since he was five years old. When it came time to attend college, there seemed to be no doubt what the Charlotte, North Carolina, native would study.

"I got a journalism degree," he said.

But, spoiler alert, Ed did not become a journalist. Instead, he stuck with what he always knew, singing and dancing, and he took those talents to the Walt Disney World Resort.

"I was part of *Beauty and the Beast* when the show opened at Disney–MGM Studios [now Disney's Hollywood Studios] in 1991," he said.

"Then came a call from Disney Cruise Line and a spot in the very first *Disney Magic* cast of performers."

Soon after that, Ed went from singing and dancing to choreographing, directing, and overseeing operations at the Disney Cruise Line live entertainment facility.

"I'm very proud of that facility and what we've done," he said. "We're turning over casts every six months, so that's currently eight casts with even more to come."

Carlos Jimenez Managing Producer, Entertainment, Disney Cruise Line

Carlos Jimenez has come a long way since his first job at the Walt Disney World Resort.

"I sewed names on the back of Mickey Mouse ears," he said of working in the Magic Kingdom when he was still in high school.

Today, he's responsible for leading the development and execution of new entertainment experiences for Disney Cruise Line.

Carlos had seemingly been preparing for this role his whole life, through his love of both Disney and entertainment. "Disney was definitely part of my childhood," said the Jersey City, New Jersey, native. "I remember dancing around to the *Mickey Mouse Disco* vinyl album and visiting Walt Disney World for the first time when I was five." After his family moved to Florida, he gravitated toward the stage and theater, doing everything from acting to stage managing for various shows and events.

"It was always a goal to come back to the magic," he said. In 2009, he did just that, working through a variety of roles at Walt Disney World Resort before taking on his current position with Disney Cruise Line.

The first-ever Disney attraction at sea, AquaMouse is a waterslide with a visual twist. Before being propelled through 760 exhilarating feet of snaking tubes with breathtaking views of the sea, riders are immersed in a cartoon world starring Mickey and Minnie Mouse.

It's a Bird! It's a Plane! It's AquaMouse!

The upper decks of the *Disney Wish* are awash with water—with more watery escapes here than on any other Disney cruise ship. But don't be alarmed. "The new pool deck, with those layers, offers multiple pools and has twice as much water as we've ever had," said Mo Landry. "Pools are so important with kids. They don't care how big it is. They just want to get wet and blow bubbles and splash around."

To that end this vessel features seven family pools and a play area spread over three decks. Plus, there's the Quiet Cove Pool, which, as the name suggests, is an oasis for adults only who are seeking a calm watery spot far from the splashing crowds.

The family pools are named for the classic Disney characters, ranging from Mickey's Pool, which fronts the Deck Stage and Funnel Vision on Deck 11, to Daisy's and Pluto's splash pools on Deck 12.

The Toy Story Splash Zone on Deck 12 is a family play area inspired by the Disney•Pixar short *Partysaurus Rex*, in which the timid tyrannosaurus rex dinosaur toy accidentally becomes the king of the bathtub. It features Woody, Buzz Lightyear, and a number of other characters from Disney•Pixar's Toy Story movies series designed as colorful bath toys. There's also a Slide-a-saurus Rex in the Splash Zone, a double-looping family waterslide.

But above them all—quite literally—is AquaMouse.

And it's not just a waterslide; it's the first-ever Disney attraction at sea, featuring powerful water jets that propel two-person rafts through 760 feet of snaking tubes that offer breathtaking views of the ocean and the *Disney Wish* itself below before splashing partakers down into a lazy river at the end of a plunge.

"We've never done anything like this before," said Danny Handke, creative director, Walt Disney Imagineering. "This is at the level of what you'd find in our theme parks: storytelling, getting wet, going fast. The overall experience just hits all the senses."

The story for people hurtling by is that Mickey and Minnie, noting the popularity of Port Adventures, have decided to start their own version of the seaside excursions—right on the *Disney Wish*.

"There's a hand-built feel to their operation," said Handke, "as if everything has been cobbled together somewhat haphazardly; but, of course, it hasn't."

Lauren Fong, assistant project manager, Walt Disney Imagineering, backed that interpretation up. "Doing AquaMouse is like doing an attraction," she said. "It has all the classic elements of Imagineering. We've got ride, we've got show, although I have to say that we've never done anything at this scale on a cruise ship."

The experience is like being in a Mickey Mouse animated short (actually, there are two different cartoons), which guests see on nine porthole screens as they traverse the "up" ramp.

"You board a ride vehicle and go through this wild water adventure with Mickey and Minnie," said Handke. "There are water effects; there are lighting effects; there's an awesome music track and show scenes. It's going to be such a memorable experience for the whole family—and you're going to get soaking wet."

Nine animated portholes and all-encompassing lighting, audio, and water effects plunge guests into the cartoon realms of either "Scuba Scramble" or "Swiss Meltdown."

Installation of the tubes for AquaMouse began while the gigablock was stationed outside the Meyer Werft construction hall under the clear, blue skies (well, a few days) of Papenburg, Germany, during the summer of 2021.

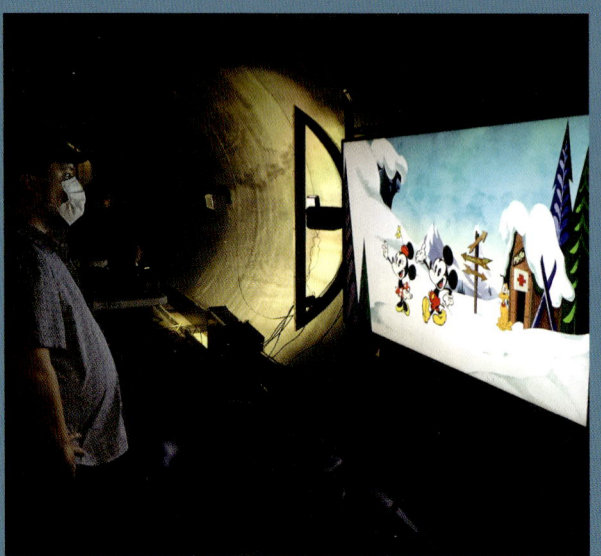

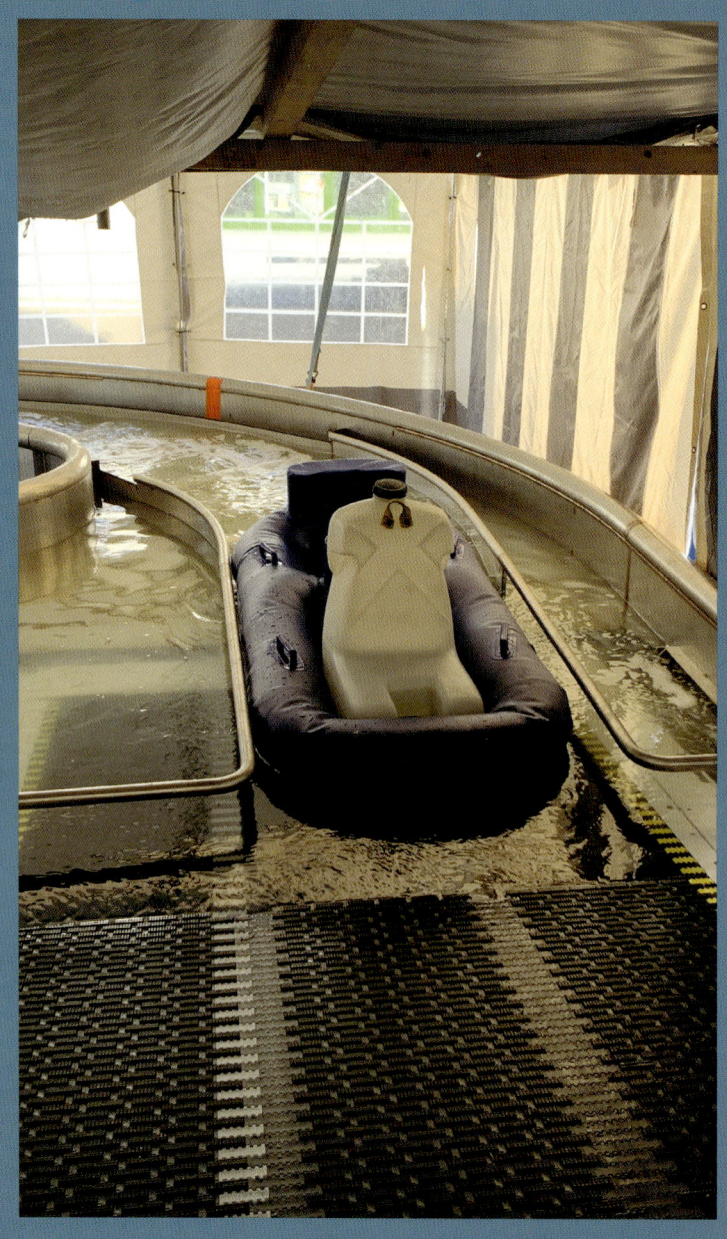

Laura Cabo (far left top), Sachi Handke, and team review plans for AquaMouse in the Digital Immersive Showroom (DISH). A test "dummy" (left) takes the plunge at a construction facility in Rasdorf, Germany. Danny Handke reviews footage of "Swiss Meltdown" in a mockup of the conveyance tube at the start of the attraction (far left bottom).

Back inside the construction hall at Meyer Werft, installation of AquaMouse continued at a feverish pace. The ascending tube in the background houses the animated sequences and effects for the attraction.

Lauren Fong

Assistant Project Manager, Walt Disney Imagineering

Blame it on Lauren Fong's parents.

"They started it," she said. "They honeymooned at Walt Disney World and we vacationed at Disneyland every year."

But as much as Lauren loved Disney growing up, and as "cool" as she thought it would be to work at Disney, "Honestly, performing on *Saturday Night Live* or being an astronaut would be more realistic than becoming an Imagineer."

Instead, she earned a bachelor's degree in Earth and Environmental Engineering from Columbia University, participated in a U.S. Department of State program in Germany called the Congress–Bundestag Youth Exchange for Young Professionals, and received a master's degree in Industrial Ecology from Erasmus Mundus, a program sponsored by the European Union that enabled her to study in Europe and Bangkok, Thailand.

But none of that seemingly prepared her for a career as an Imagineer, though she did learn German. "I went there [Germany] with absolutely nothing and by the end of two months I was talking like a toddler," she said. 'I'm hungry.' 'I need that.' 'I want that.' 'I'm sleepy.' I had it down."

Eventually, Lauren did manage to land a project management internship at Walt Disney Imagineering—Florida in January 2015 working on an expansion of Toy Story Mania!

"Do what you love, not what you think Imagineering wants," she said. "You'll feel satisfied with yourself and I'm testament to that."

In 2017, shortly after taking a full-time position with Imagineering, she went to the Netherlands for an Erasmus Mundus reunion. "I told everyone that I never knew if I was going to be in Europe again," Lauren said. "I have a job now and I can't just come to Europe for two weeks every year because I also want to see my family in California. So I said goodbye to everyone."

Then, on her first day back at work after returning home, she received an email.

"It was from someone who heard that I might be interested in working on the cruise ship project, which is funny because I had never expressed an interest in the project, even though it's something I would really be interested in."

Not surprisingly, Lauren jumped at the opportunity, even though she later learned she wasn't exactly at the top of the team's list of candidates.

"I'd like to think that I was the first choice, but I think it came down to the fact that I speak German," she said. "Trevor Noah wasn't the first choice for *The Daily Show* either, but if I'm the Trevor Noah of the *Disney Wish* project, I'll take it."

Far from the madding crowds of AquaMouse, Toy Story Splash Zone!, and the family pools is the Quiet Cove, an adults-only oasis on the back of the ship with in infinity pool, whirlpool spas, and endless views of the sea.

In *Star Wars*: Cargo Bay, kids become a new generation of tech and mechs supporting the Resistance. They'll get to build droids, manage creatures and meet a few of their *Star Wars* heroes.

The Whole *Youth* and Nothing but the *Youth*

Kids are the focal point on Disney cruise ships—and the *Disney Wish* certainly continues that tradition with an array of youth-*centric* spaces, activities, and entertainment opportunities that have jealous adults wishing they were young again. In inventive spaces designed exclusively for them, youngsters, depending on their age, can take part in a wealth of imaginative experiences and activities (guided by specially trained Disney counselors-at-sea) or just kick back in hangouts with others their age.

One of the first things Disney cruise veterans will notice is that the "it's a small world" nursery and Disney's Oceaneer Club have been consolidated into one space on Deck 2.

"Back when we started with the *Disney Magic*," said Mo Landry, "we had very distinct age groupings with specific programs for each age group.

"For instance," she added, "all of our princess programs back then were in the three- to four-years-old age group. Well . . . not every three- and four-year-old liked princesses and . . . , [we found] there were older kids who loved princesses."

The solution was to bring all the youth activities areas together into one place and combine them with a more flexible design. "By making that adjustment," Landry said, "turning the area into a place that can accommodate three- to twelve-year-olds, the kids get to choose what they want to do.

"Kids don't need to be programmed on their vacation. This is their time to pick and choose anything they want to do. They want to go be a pirate, they can be a pirate. They want to go be a princess, they can be a princess. It widens that lens of what they can do and also, from an inclusivity standpoint, it lets them choose what they like instead of putting them into a specific box."

Part of the new design also included a simple but far-reaching modification: hallway access doors to each activity area.

"It was an idea that came from a crew member," explained Landry. "On the *Disney Fantasy*, we took Andy's Room and put a door to the hallway so that we could use that space for toddlers at certain times of the day. It worked incredibly well.

"Now, on the *Disney Wish*, we've expanded that concept to most of the rooms," she said. "They can be shut off from the rest of the Oceaneer Club and opened to families to do a program or to just come in and play. For instance, we can bring teens and tweens, or even adults, into *Star Wars*: Cargo Bay, but we don't have to close down the rest of the Oceaneer Club to do it. We have the best of all worlds now and it's a big win for us and our guests to have that flexibility."

Naturally, that begs this question: What are in the youth activities spaces that would prompt everyone to want to be there?

The short answer? A lot.

Chronologically, it starts with the "it's a small world" nursery, which is exclusively for children from six months to three years old. As the name suggests, the design of the nursery takes it cues from the classic Disney park attraction. The area, which includes a nap room and a play space, features a colorful patchwork of buildings, geometric shapes, and painted landscapes of Disney, Disney•Pixar, and Marvel characters and stories from around the globe . . . and beyond. The whimsical artwork was created by artist Joey Chou, channeling Disney Legend animator and artist Mary Blair, whose distinct style set the tone for the original "it's a small world" attraction at Disneyland.

The light and airy nature of the nursery even extends to the floor, which is in varying shades of blue to represent the attraction's water flume.

As for Disney's Oceaneer Club, where can one truly begin? Yeah, there's that much there.

Although the club's on Deck 2, parents can check in their children in the Grand Hall on Deck 3 and then send them "down the rabbit hole" (a twisting slide) directly and securely into the space. The check-in area is themed to *Alice in Wonderland*, rendered in a colorful, childlike style that, like the nursery, recalls the designs of Disney artist Mary Blair.

At the center of Disney's Oceaneer Club is the Hub, which serves as a portal to the other areas of the facility and features a stage for specials events, interactive storytelling, and in-person and virtual visits from Disney characters.

The "it's a small world" nursery, for children six months to three years, was inspired by the classic Disney park attraction. It features (clockwise from top) murals by artist Joey Chou (channeling Disney Legend Mary Blair) and a number of inactive elements featuring such familiar Disney icons and characters as Peter Pan, Wendy, John and Michael; a whimsical train with Mickey and Minnie aboard that "steams" around the nursery's perimeter; cutouts of Marie and a can-can dancer from *The Aristocats*; and the palace of Agrabah.

Mickey and Minnie Captain's Deck (center left) is actually part of Disney's Oceaneer Club, but the babies and toddlers in the nursery will have supervised access to that space, too.

Disney's Oceaneer Club, always a popular place for children 3 to 12 on Disney Cruise Line ships, offers opportunities for storytelling, dress-up, creativity, arts and crafts, toys and games, and even visits from favorite Disney, Pixar, Marvel, and *Star Wars* characters. In Fairytale Hall, Rapunzel's Art Studio (top) is the setting for arts and crafts, while Belle's Library (center left) hosts story time and Anna & Elsa's Sommerhus (center right) is the place for gesture-based, interactive "snow" games. Kids suit up for action at Marvel Super Hero Academy (bottom left), while the Walt Disney Imagineering Lab (bottom right) gives them the chance to design and ride their own roller coaster.

Then, it's off to Mickey & Minnie Captain's Deck, a maritime-themed space with physical and sensory-style games and activities for the youngest kids; Marvel Super Hero Academy, where would-be Avengers suit up and use gesture-based technology to battle the bad guys; *Star Wars*: Cargo Bay, in which techs and mechs work behind the scenes to make sure the Resistance's operation runs smoothly and efficiently; and Walt Disney Imagineering Lab, where kids learn about Imagineering and get their chance to try their hand at being an Imagineer.

There's also Fairytale Hall, featuring a trio of activity rooms centered around Disney royalty. In Belle's Library, it's story time, where favorite tales are read and acted out with help from Disney friends. At Anna & Elsa's Sommerhus, snow is never far away, even in this cozy summer cabin where Olaf hosts "Frozen Fun." Here, kids use their newly found ice powers to play several gesture-based games. In Rapunzel's Art Studio, meanwhile, creativity is key as youngsters work on their own crafts projects (maybe a floating lantern?) and perhaps paint a masterpiece (or at least draw a Disney character or two).

And that brings us to the tweens and teens, who have come a long way from the first Disney ships, when they got a bit of the short end of the stick.

"Teen and tween spaces and programs have evolved over the years," said Mo Landry. "On the *Dream* and the *Fantasy*, we finally actually designed spaces intended for them, as opposed to trying to adapt existing spaces, which we did on the *Magic* and the *Wonder*.

"On the *Wish*, we've taken it to another level," she continued. "What we have created are spaces that take into account everything we've learned. For instance, the tweens [approximately ages eleven through fourteen] have their own custom-built space with a secret entrance and their own elevator going to the upper decks."

The name hasn't changed for the tween space, but the look and feel of Edge have. The Deck 4 club has taken on the appearance of a New York City loft, outfitted with items that might seem more at home at a Central Park picnic, including furniture covered in fabric designed to look like picnic blankets, poufs shaped as tree stumps, and carpet that resembles grass lawns with daisies.

And about that secret entrance and dedicated elevator, tweens are not only the only ones allowed in; but they are the only ones who know how to get in.

As for the teens (ages fifteen to seventeen), "They always want to be in the next age group," said Landry. "They want to be eighteen. They want to be able to go to the clubs. They want to go to Enchanté and PALO Steakhouse. The space that we have designed for them [called Vibe] is truly raising the bar, giving them beautiful views, their games, special coffee machines, and just a place where they can be comfortable hanging out with their friends."

Vibe has the look of a classic Parisian loft overlaid with an eclectic retro twist. Open and spacious, with natural light pouring in from floor-to-ceiling windows, the space and its traditional architectural elements come alive with Disney-inspired, paint-drenched columns, hand-painted wall murals, and contemporary graphic art.

Adjacent to Vibe on Deck 12 is The Hideaway, which serves as a flex space for teens, tweens, and young adults. As with so many of the youth areas, this one might seem attractive to parents as well, with its retro designs and murals inspired by the 1977 revival of *The Mickey Mouse Club*. And that's okay, because it also gets used for family events.

The evolution of the teen and tween spaces over the years has been successful in at least one respect: "Those kids really get to know each other in such a short time," said Landry. "At the end of a three- or four-night cruise, you get tears, lots of them."

The Hideaway, adjacent to Vibe on Deck 12, is a flex space for teens, tweens, and young adults that incorporates retro designs and murals inspired by the 1977 revival of The Mickey Mouse Club.

144   Making a *Wish*

Inspired by the look of a classic Parisian loft overlaid with an eclectic retro twist, Vibe (top) is the hangout for teens, while Edge (bottom), for pre-teens, takes on the look of a New York City loft decorated with items that might seem more at home at a picnic in Central Park (and has its own secret elevator entrance).

Captain Minnie Mouse

Who's the leader of the club—and classic cartoons, company, and cruise ship—that's made for you and me?

In the past, the answer would no doubt be Mickey Mouse. That's the way it's seemingly always been, even before Walt Disney led off his very first episode of the *Disneyland* television series by declaring that "I only hope that we never lose sight of one thing—that it was all started by a mouse." (And, by mouse, he meant Mickey, whose name was on the cover of the book Walt had in front of him. If there were ever any doubt, Walt went on to say, "Now that's why I want this part of the show to belong to Mickey.")

Well, on the *Disney Wish*, Minnie Mouse is no longer playing second fiddle to that other mouse.

On this ship, she's the captain (shh, don't tell Marco Nogara). Her face is on the bow, she presides over the toddler play area in Disney's Oceaneer Club, and, when guests dive into Disney Uncharted Adventure, they'll discover that it's Minnie—move over, Mickey—who's the one running things.

It's about time, really. After years of standing to Mickey's side while everyone celebrates his birthday on November 18 (it's her birthday, too, by the way), seeing Mickey stand next to Walt in front of Disney park castles (she does get to sit by Roy, Walt's brother and business partner, in a few of them), and hearing nothing but "Mickey, Mickey, Mickey" like Jan does "Marcia, Marcia, Marcia" in *The Brady Bunch*, Minnie is finally stepping out in a crisp new uniform with smart white trousers and a bold red jacket emblazoned with captain's insignia.

Minnie's long-overdue promotion is no accident. The debut of her career as a captain is intended to inspire the next generation of female leaders in the maritime industry, empowering girls and young women to chart a course for success in a field they may not have thought about before this.

In addition to seeing Minnie onboard the *Disney Wish* in her new role (guests may already have seen her as a captain-in-training on the four existing Disney ships), children can transform into seafaring captains with a Captain Minnie Mouse makeover at Bibbidi Bobbidi Boutique, available not just on the *Disney Wish*, but every Disney ship.

So, in the future, when you hear that it was all started by a mouse, know that it wasn't just Mickey. Just as importantly, it was Minnie, too.

Imagineers Ken Horii (left) and Manny Ramirez discuss an interactive screen that will be part of the Marvel Super Hero Academy experience.

Staterooms on the *Disney Wish* are airy, spacious, comfortable, well-appointed—and feature artwork from Disney animated classics above the headboards and throughout the rest of the room.

The State of the Rooms

As enchanting as the public spaces are on the *Disney Wish*, perhaps the most important place is the one in which guests spend the most time: their staterooms (except for a teenager, who probably prefers to spend every waking—and sleeping—hour in Vibe). Disney Cruise Line ships already feature standard staterooms that are among the most spacious, comfortable, and well designed in the industry. The *Disney Wish* takes that one step further with accommodations infused with Disney stories that build on the ship's theme of enchantment.

"The rooms are essentially the same throughout the ship," said Claire Weiss. "It's just the artwork that's different from room-to-room, which is different from the staterooms on our existing ships. It gives guests the element of surprise; and that's part of the fun of it."

The most prominent piece is the headboard above the bed, which features a background from the animated film featured in the room.

"It's the 'establishing shot' right when you walk in," said Weiss. "It's the highlight of each room and has the most impact on the room's tone and character. We've put the emphasis on color and composition, infusing it with a little Disney magic in the form of gold and silver leaf on each piece that is meant to draw attention to various enchanted icons."

Artwork in the rooms ranges from the fairy-tale castles of *Cinderella* and *Frozen*, to the enchanted forests and animals found in *Sleeping Beauty* and *The Princess and the Frog*, to the spirit of the sea, embodied in *The Little Mermaid* and *Moana*.

"My favorite is the *Moana* [headboard] piece," said Weiss. "It was digitally painted by James Finch, who was one of the visual development artists for the film. We actually sat with that team and explained what we wanted and then they came back with some rough sketches. During that whole process, I kept telling myself, *I can't believe I'm telling James Finch what to paint, but also still do it, please.* It was one of those moments, and there are many, when I can't believe that this is the job that I have."

Each stateroom also has at least one smaller piece over the sofa, usually featuring two main characters from the animated film the room is centered on. Some rooms even have a third piece, a "close-up" by the vanity, that focuses on loosely drawn "sidekicks," often lovable and/or adorable characters that bring warmth and humor to the story.

"I really love the work that the staterooms have undergone because that was a lot of collaboration by a lot of different people," said Weiss. "I think the artwork is stunningly beautiful. All of those pieces were either created for us or modified for us, but they're all sort of unique to the *Wish*."

There are a lot of understated design elements throughout the staterooms, but Weiss points to one in particular. "One of the unexpected things we have in the staterooms is the design that's in the lampshade," she said. "It's such a subtle thing. The backing of the lampshade actually has a cutout that has filigree from the sheets with Mickey Mouse's head. When the lamp is off, it's just a lamp. When you turn it on, Mickey's head comes through the shade.

"We talked a lot about how to get 'Disney' in these rooms without it feeling like overkill," she continued, "so that felt like the exact perfect blend of 'Disney' and 'nice.' It's so refined and it's just unexpected. It's a sweet little magical thing."

Standard staterooms make up the bulk of the accommodations aboard the *Disney Wish*, but as demand has grown for the "concierge experience," so has the number of suites. The number of concierge-level rooms on the ship nearly doubles that of the *Disney Fantasy*, increasing it to seventy-six (from forty-one).

All the suites draw on Disney's *Tangled* for inspiration: the moment in the film when thousands of colorful lanterns rise above the castle and cast their reflections in the water while Rapunzel and Flynn Rider look on from a boat in the harbor, for example, is illustrated in the headboard artwork. This scene also informs the color palette of the rooms.

Disney Wish also features a first for a Disney Cruise Line ship: seven forward-facing suites above the bridge that offer full frontal views of the ocean through floor-to-ceiling windows that overlook the bow. These suites find inspiration in Disney's *The Little Mermaid* for artwork, carpeting, and light fixture designs.

But we're not done yet.

In another first for the Disney fleet, there are four "royal" suites that celebrate the enchanted world of *Sleeping Beauty*, with two Princess Aurora Royal Suites and two Briar Rose Royal Suites beckoning. Each pair includes two-master bedrooms, each with its own bathroom, in a single-floor option and a two-story configuration.

CONCIERGE SUITES

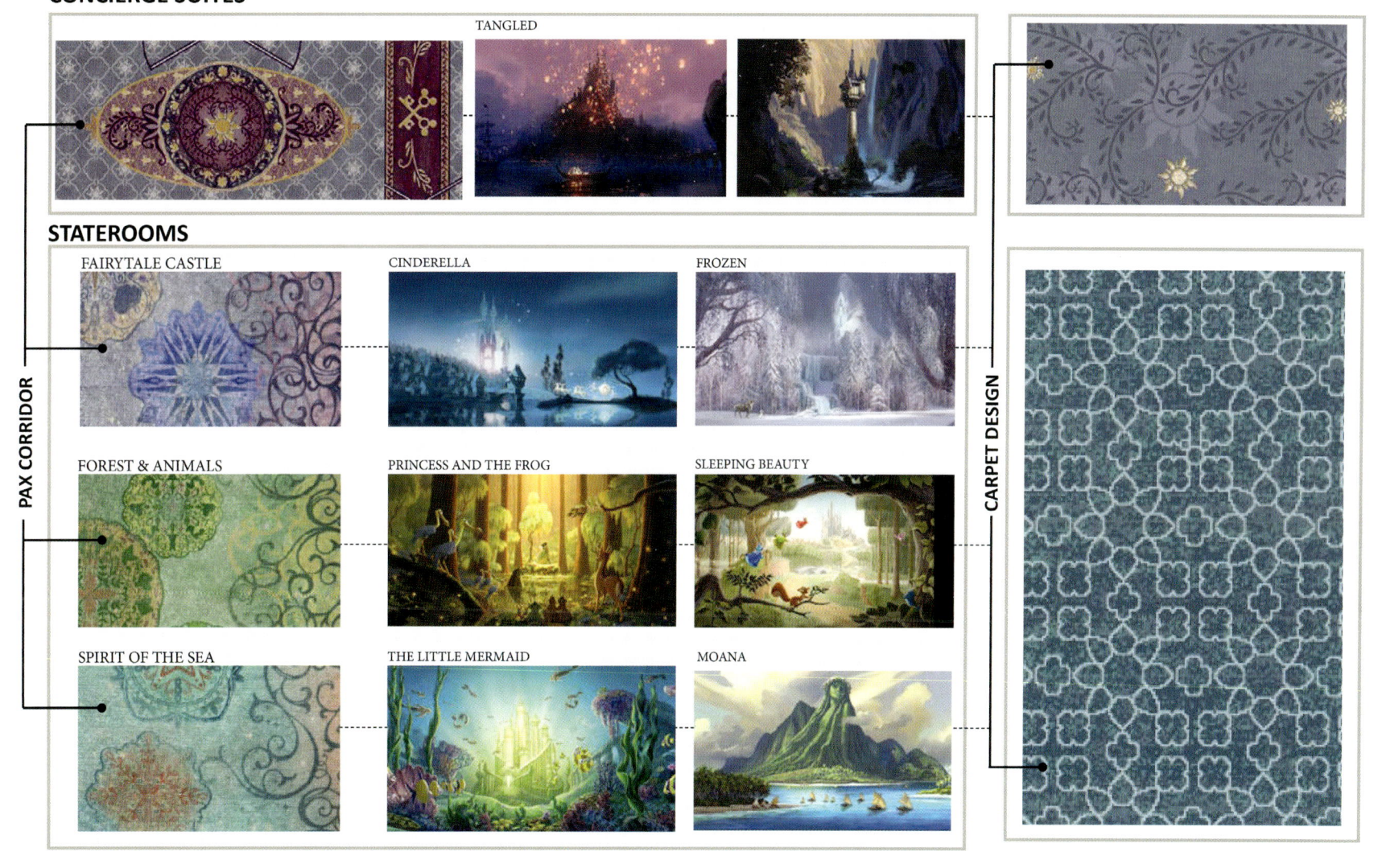

TANGLED

STATEROOMS

PAX CORRIDOR

FAIRYTALE CASTLE	CINDERELLA	FROZEN
FOREST & ANIMALS	PRINCESS AND THE FROG	SLEEPING BEAUTY
SPIRIT OF THE SEA	THE LITTLE MERMAID	MOANA

CARPET DESIGN

Staterooms boast most of the same functional elements—sofas, carpets, draperies, cabinets—with one very important design difference: staterooms feature artwork inspired by various Disney animated films revolving around the theme of enchantment.

In addition to the large "establishing shot" above the headboard, staterooms also contain secondary pieces that tie the space together. This art features main characters and, in some rooms, loosely rendered "sidekicks," including Flounder and Sebastian from *The Little Mermaid* (top left) and Olaf from *Frozen* (bottom left), as well as Moana (top right), Briar Rose and her woodland friends (center right), and Princess Tiana (bottom right).

Workers install the art above the headboard for a "Cinderella" stateroom (don't worry, Olaf will go in a *Frozen* room) (top). Lampshades offer a magical little prize when the light is turned on (bottom left). Staterooms were assembled offsite and then craned onto the ship and rolled into place (bottom right).

Suites, which accommodate up to five guests in suite and one-bedroom configurations, draw on *Tangled* for inspiration (above). New to the Disney Cruise Line fleet are forward-view concierge suites (opposite), which offer full-frontal views of the sea ahead and artwork from *The Little Mermaid*.

All suites offer access to exclusive areas and amenities, including the Concierge Lounge, which is more than triple the size of similar lounges on the existing Disney ships. It also boasts an unbeatable location on the ship that includes an indoor lounge, an outdoor terrace, and a private sundeck with two whirlpools and unobstructed—and stunning—views of the sea.

Which brings us to the height of opulence on any Disney ship: the Wish Tower Suite.

Housed in the forward funnel above the upper decks and with Disney's *Moana* as its design inspiration, the two-story accommodations include two master bedrooms and a children's room with bunk beds on Deck 15, plus a library that can also serve as an additional en-suite king bedroom on Deck 14. The double-height living room features a floor-to-ceiling glass window that offers a view of the sea.

And, yes, it has its own private elevator, just in case you were on the fence about whether you wanted to book it or not—and feared having to navigate stairs when going from one floor to another.

The Princess Aurora and Briar Rose Royal Suites are two-bedroom suites (each in both one- and two-story configurations) that evoke her life as Aurora in the castle and Briar Rose in the woods. The Princess Aurora Royal Suite has a soft dream-like palette of blues and golds (the bathroom, top, and the bedroom, bottom right), while the Briar Rose Royal Suite leans toward greens, burgundy and dark wood tones (the living room, bottom left).

The Wish Tower Suite is an opulent, two-story space perched in the forward funnel. The spirit of *Moana* and the Heart of Te Fiti provide inspiration for the suite, which has two master bedrooms, a children's room with bunk beds, and a library, not to mention a double-height living room (opposite) with a floor-to-ceiling glass window that offers an expansive view of the sea.

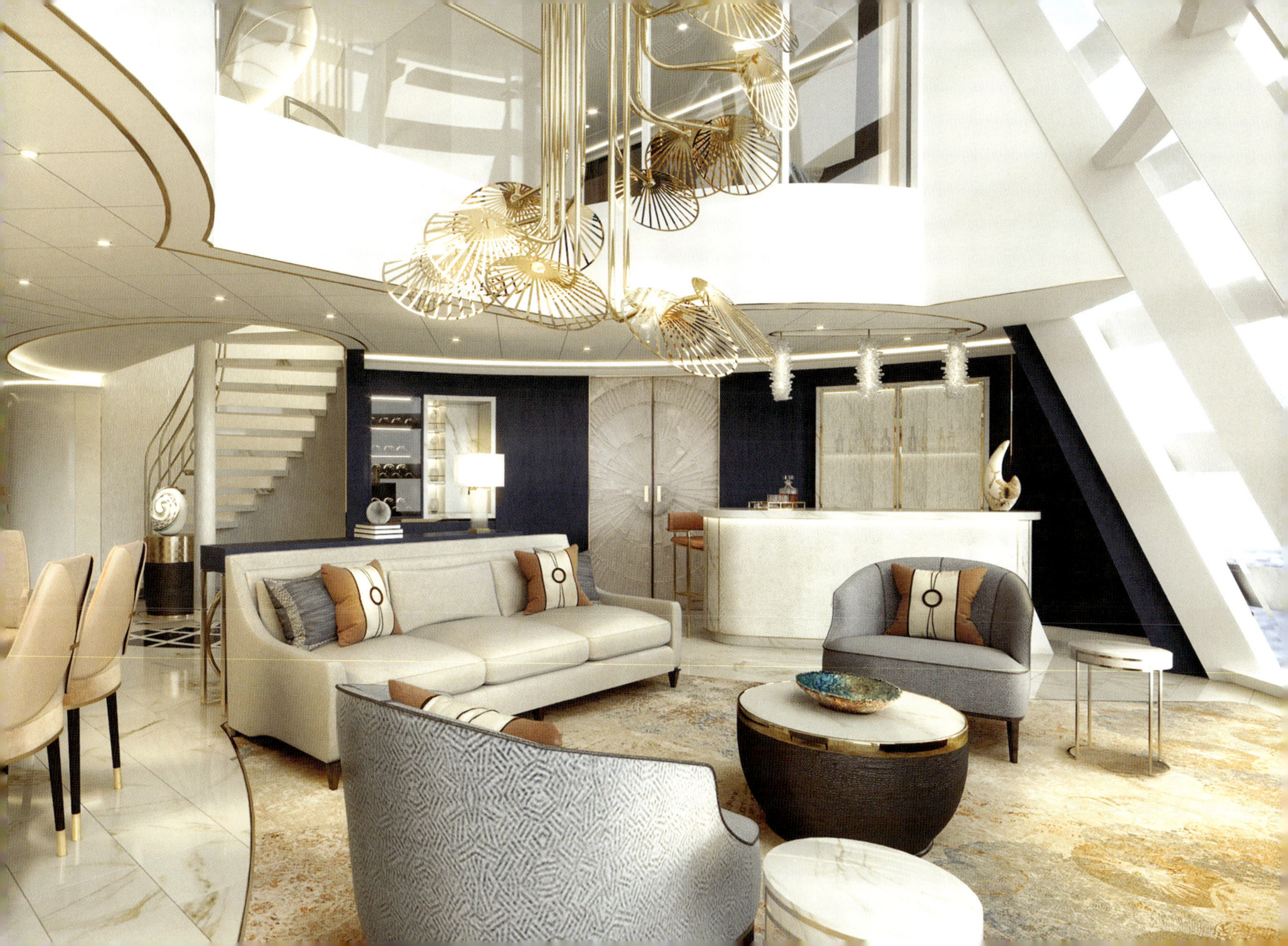

Lloyd Machado   Hotel Director, Disney Cruise Line

Lloyd Machado's first brush with Disney was *The Jungle Book*. It certainly wouldn't be his last.

"Disney was just entering the Indian market when I was growing up," he said, "and that movie got me hooked." Although he always wanted to be in the hospitality business, it would be some time before he encountered Disney again, even though his first job was on a cruise ship.

"Disney wasn't in the cruise industry at that time," he said, so he worked on a number of cruise ships—and then he met someone from Disney Cruise Line.

"It was a stroke of fate all the way," he said.

Lloyd joined Disney as a server on the *Disney Magic* in 1998. Since that time, he has practically done it all when it comes to food service, advancing to head server, restaurant manager, dining room manager, food and beverage service manager, and, today, something a little bit different, hotel director on the *Disney Wish*.

When he's not aboard a Disney ship, he lives in India with his wife, Priti, and daughter Kaitlyn.

Lisa Marie Picket   Hotel Director, Disney Cruise Line

Lisa Marie Picket has gone from milking cows as a kid in London, Ontario, Canada, to leading the hotel operations team aboard the *Disney Wish*.

"I grew up on a dairy farm," she said, "and every Sunday my family would watch *The Wonderful World of Disney*. It was the only TV we ever watched."

In college, she studied Travel and Tourism because she wanted to see the world and learn about other cultures, which she did . . . sort of. After graduation, she was hired into the Walt Disney World Resort International program.

"When I first saw Cinderella Castle in the Magic Kingdom," she said, "I could not believe that after years of watching *The Wonderful World of Disney*, here I was standing in front of a Disney castle, just like on the show."

In 1998, she joined Disney Cruise Line as part of the opening team for the *Disney Magic* and has been with Disney Cruise Line ever since. "I did not set out to become a hotel director when I graduated from college," she said, "but with the incredible experiences I've had at Disney Cruise Line, it turned out to be a natural growth path."

Maureen Landry-Mancini

Director, Entertainment Operations and Special Projects, Disney Cruise Line

Maureen Landry-Mancini—Mo, for short—is from Hudson, Massachusetts, about an hour from Boston, but it seems as if her heart was always somewhere else when she was growing up.

"My family vacationed at Walt Disney World every year," she said. "It really was our only vacation place. My parents tried to take me out to San Francisco once and it was terrible because it wasn't Walt Disney World."

Even when she was going to Northeastern University in Boston, majoring in recreation and sport management, she always seemed to find herself in and around Walt Disney World.

"I found out about the Disney College Program from a server at the Grand Floridan Café when we were on one of our vacations, and I was lucky enough to be chosen to come down after my freshman year in college because they needed lifeguards," she said. "Normally they only took upperclassmen."

Her final internship in college? Central Florida.

After an internship with the Orlando Magic—"It was during the glory days of Shaq and Penny and Horace and Nick, and it was amazing," she said—followed by a stint with the town of Celebration, Florida, as the Parks and Recreation manager, the opportunity to work at Disney again finally came.

"Disney Cruise Line needed people for the *Disney Magic* who knew kids and entertainment and activities," said Mo, "It was perfect for me."

And she definitely knew what she was doing.

"We had a number of team members who came from within the cruise industry and then we had the Disney group, and we were finding our way together," she said. "One of our dining managers came from another cruise line and where he came from the dining room is the dining room. That's all you do there. When I told him I was going to be bringing kids into Animator's Palette so that they could learn animation, he thought I was a crazy person. 'Children do not belong in there,' he said. I said, 'Yes, they do. This is a perfect setting.' We worked it out. Now, people who came from within the industry and are still with us have Disney in their soul. They've stayed with Disney because of that love."

By the way, the transatlantic crossing that brought the *Disney Magic* to the United States from Italy? That was the first time Mo had ever been on a cruise ship.

She's been on them more than a few times since then. And if she seems to do odd things, there's a reason.

"Disney was part of my life growing up, one-hundred percent," she said. "It's part of my core and I will still pick up trash in random places because that's what we do."

Artist Nikkolas Smith (a former Imagineer) offers a fresh perspective on the story of Cinderella in Cindys, one of a number of pieces of original art commissioned for the *Disney Wish*.

A Dream Is a Wish Your *Art* Makes

Disney art—sketches, renderings, backgrounds, animation cels—has always been a big part of the design sensibility of Disney cruise ships. *Disney Wish* takes this to a whole new level with more than four thousand pieces of art, much of it created expressly for the ship. "Our artwork approach is a little bit different than what we've done on the existing ships," said Claire Weiss. "On the existing ships, it's really about more vintage artwork and photographs of Walt. Here, we're using vintage, archival artwork because that is always beautiful, but we're also telling newer stories that are made in a completely different way that don't have concept artwork like you would have from a film [done] in 1935."

"We're using the opportunity to present things in a more inclusive and broader way," added Laura Cabo. "It's not just taking the wonderful Cinderella of the original film; it's having artists take on the story themselves and provide an exciting and fresh perspective with, for instance, multigenerational and multiethnic backgrounds."

The team also worked with the Disney Animation Research Library, Pixar Animation Studios, Marvel, and Lucasfilm to curate and produce art that appears all over the ship—in restaurants, bars and lounges, staterooms. Yet according to Cabo, one of the best places to see this array of art is in the stair halls fore and aft.

"Not only do they help you find your way," she said, "they're an opportunity to present our stories in a beautiful and emotional way."

Pieces in the forward stair halls focus on moments of transformation, usually involving characters, such as Maleficent from *Sleeping Beauty* and Rapunzel and Flynn from *Tangled*. The aft stair halls focus on the fantastical locations highlighted in Disney films. They're often (but not always) castles, such as Arendelle from *Frozen* and the bayou from *The Princess and the Frog*. (To that end, in keeping with Cabo's emphasis on helping guests find their way around the ship—a desire of Cabo's that arose from her own experience aboard a Disney cruise—stair halls, elevator landings, and stateroom corridors have been coordinated, in design and color, to provide subtle wayfinding clues that give guests a better idea of where they are on the ship.)

There are other forms of "art" on the *Disney Wish* as well, found in the design of carpets, wall coverings, railings, fixtures, porthole frames, and even ceilings.

And then there are props.

"Props are an additional overlay of storytelling," said Kristen Zeigler. "What characters live in this space? What items might they have that are populating it? It's anything that helps enhance the space and tell the story."

Part of Zeigler's job does involve treasure hunting, but there's more to it than that. "A lot of people think you just go down to the antique store or a thrift store, or eBay, buy a vase, and plop it on a shelf," she said. "But even the real-world things that we buy, we have to do a lot of modification work to make them 'guest ready.'"

And then there are the items that don't even exist in the real world.

"About fifty percent is custom fabrication," Zeigler said, "especially if we're trying to replicate something from an animated film. You just can't go buy it in real life because it doesn't exist and never did, so we work with our partners, whether that's Pixar or Disney Animation or Marvel, to sort of re-create these items; except we're doing it for the first time."

Zeigler also works with artisans from around the world to fabricate the "authentic" props she needs.

"One example is I worked with a couple in Finland who custom fabricate drums," Zeigler said. "They're creating a sami drum for the Northuldra people from *Frozen II* that will go into 1923. It's not a huge showcase prop, but it shows the authenticity of the process we put into every single piece."

She's also working with an artist in Hawaii who does custom ship models. "He's creating a model ship, also for 1923, that is based off the Fijian ship that was the inspiration for Moana's boat."

And then there are the clocks.

"We've custom-fabricated several clocks for the ship," said Zeigler. "There's a clock that you see in Rapunzel's tower in *Tangled* that we've built for Rapunzel's Art Studio in Disney's Oceaneer Club, and a clock you don't see in *Frozen*, because it's not actually in the film, but it is inspired by it. We've put that one in Anna & Elsa's Sommerhus [also] in Disney's Oceaneer Club.

"And then we do have the iconic grandfather clock from *Frozen*," she said, "the one where Anna is doing the 'ticktock, ticktock' in front of it; that

ELEGANT RESTAURANT PROPS

Props provide an additional layer of storytelling, whether it's in the form of sketches, background art, maquettes, and art supplies placed in displays to make 1923 feel as if the restaurant is a Hollywood animation studio (top) or "authentic" pieces, such as a sami drum from *Frozen II* (bottom left) or Moana's boat (bottom right), that had to be manufactured from scratch because, as props from an animated film, they never existed before in the real world.

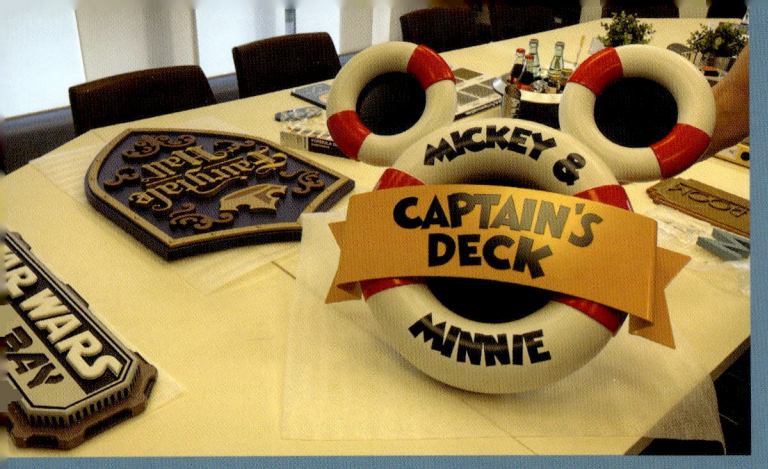

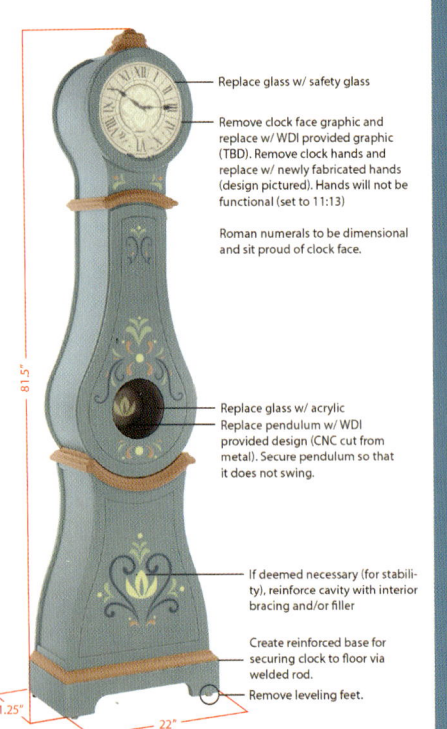

Signs and graphics serve both a thematic and practical purpose, providing vital information in an artistically pleasing way. Samples of a few of the signs found in Disney's Oceaneer Club (top), the development of the logo for Enchanté (center right, the final version is opposite), and graphics for stateroom numbers and hallway light fixtures (bottom right). Back to props, an Arendelle-inspired clock design—and the actual clock, which can be found in Anna & Elsa's Sommerhus in Disney's Oceaneer Club (bottom left and center).

Clock specifications:

- Replace glass w/ safety glass
- Remove clock face graphic and replace w/ WDI provided graphic (TBD). Remove clock hands and replace w/ newly fabricated hands (design pictured). Hands will not be functional (set to 11:13)
- Roman numerals to be dimensional and sit proud of clock face.
- Replace glass w/ acrylic
- Replace pendulum w/ WDI provided design (CNC cut from metal). Secure pendulum so that it does not swing.
- If deemed necessary (for stability), reinforce cavity with interior bracing and/or filler
- Create reinforced base for securing clock to floor via welded rod.
- Remove leveling feet.

Dimensions: 81.5" × 22" × 1.25"

will be in the Arendelle restaurant." That one is the only one that is actually functional. The time on the other two, and this is a bit of an Easter egg, is set to the month and the year the films came out, so it's 11:10 for *Tangled* and 11:13 for *Frozen*."

Another type of art guests will spot on the *Disney Wish* comes in the form of graphics. "Graphic design has always been based in what was earlier called commercial art, and it's always been an art form in my eyes," said Alexis Cummins, graphic designer, Walt Disney Imagineering. "The complexity of what we're doing makes it art.

For Cummins, graphics are always an opportunity to tell a story, not just in signage and marquees, but in menus, plates, printed items, napkin embroidery, and even the "A" for Avengers that is on the cutlery in Worlds of Marvel. "It's that level of detail," she said, "what fits and what works best and what carries the story."

Cummins's most obvious contribution comes in the form of the sign that usually welcomes guests to a club, a restaurant, or any other place on the ship, for that matter.

"You're educating guests about where things are and the marquees become really a port of entry to whatever experience you're about to go into," she explained. "When you're stepping into the Hook's Barbery or Untangled Salon, that's your first impression of this new and immersive experience you're about to have. The sign gives you a clue as to what's behind that door, both from the name of the facility and the nods to what the experience might be that are incorporated into the design of the marquee."

One example is Enchanté, the premium dining experience on Deck 12.

"For Enchanté," Cummins said, "we were really careful about how that design happened. The typography evolved into a place where we could naturally introduce the character presence, so when you look at it, you say, 'Oh, I'm going to this dining experience and it's related to "Be Our Guest," and it's linking that all together in a very sophisticated way."

Of all the designs she's done on the ship, Cummins actually does have a favorite. "I really do like Nightingale's," she said. "When we were designing the space, we always liked the idea of referencing the stepsisters, referencing Cinderella," she continued. "So the typography is inspired by one of the original *Cinderella* movie posters. We also added an actual nightingale within the signage as part of the apostrophe. The bird is singing, and those musical notes are evident in the typography as well."

For every beautifully detailed marquee, though, there are plenty of operational and safety signs that had to be designed as well. "I never thought I'd be designing the side of the lifeboats," said Cummins, "and I'm doing that."

But it's that attention to detail that distinguishes the Disney ships.

The *Disney Wish* Joins the Disney Cruise Line Fleet, by Los Angeles–based artist Matteo Marjoram, can be found in the forward stairs on the landing between Decks 3 and 4.

Alexis Cummins Graphic Designer, Walt Disney Imagineering

Alexis Cummins has always had affinity for art—and Disney.

"I have photos of me at a very early age drawing Mickey and Donald," she said. "That's how I related to Disney from that early of an age."

When it came time for college, Alexis chose the University of Florida and graphic design, but she never realized that could lead to a career at Disney.

"I remember watching 'Imagineer That!' on the Disney Channel and seeing all these different creative disciplines at Walt Disney Imagineering," she said. "At that time, I didn't know graphic designers existed at Disney."

Until a group of Imagineers came to the University of Florida on a recruiting trip.

"I happened to see a flier at my college that said Disney is coming to review portfolios," she said. "I was the first person in line."

After that, things happened very quickly.

"I was lucky enough to get an internship right away in the graphics department," said Alexis, "and I've been here ever since."

Kristen Zeigler Set Decorator, Props, Walt Disney Imagineering

Kristen Zeigler grew up loving the Disney parks, so when she was assigned a book report her a freshman year in high school, there was no doubt what she was going to do.

"I knew I wanted to do mine on Walt Disney Imagineering," she said. "That book, *Walt Disney Imagineering: A Behind-the-Dreams Look at Making the Magic Real*, opened her eyes to who and what Imagineers are, so much so that working for Disney became her singular focus.

After graduating with a degree in graphic design from Cedarville University—"a small school in the cornfields of Ohio"—Kristen saw an internship posted for set decoration and props at Walt Disney Imagineering in Florida.

"I had never really done props," she said. "But graphics was my background."

She got the internship anyway. Part way through, she was offered a full-time job—in graphics. She turned it down. "I decided to stay in the props department," she said. "I loved what I was learning, and that's what brought me to where I am today."

With Disney Uncharted Adventure, guests will discover hidden Disney worlds of magic, wonder, dreams, and fantasy (you see what we did there?) on and around the *Disney Wish*. The adventure is a multi-player, multi-day role-playing game that uses mobile devices and immersive ship technology to put guests in the center of the action.

Wish Upon a Star and Save the Universe

New to the Disney Cruise Line on the *Disney Wish* is a multiplayer, multiday role-playing game using mobile devices and immersive ship technology that places guests at the center of a grand adventure through heretofore hidden worlds of Disney magic, wonder, and imagination. Wait, isn't there already something like that on the *Disney Dream* and the *Disney Fantasy*? In a way, yes, but Disney Uncharted Adventure (that's the name of the new journey) is to Midship Detective Agency what *Minecraft* is to, say *Tetris*: they're both incredibly fun games, but advances in technology have raised the bar on what can be done with interactive experiences.

And, although you or someone in your group does need a mobile device to participate, you won't necessarily have to be well versed in video game-play.

"We have a lot of technology power in this experience, but you should not be made to feel like a technology person to play it," said Davey Feder, software product manager, Walt Disney Imagineering. "Our goal with the technology is that the more it can live in the background, the better. The technology should never be the star. The technology is just a tool we have as storytellers that allows us to bring new experiences to life."

The bottom line, he said: "It shouldn't matter if you're tech-friendly or techno-phobic. You should be able to jump in and have it work for you either way."

The same is true when it comes to using your phone during the experience.

"We don't think of this as a phone experience," said Feder. "We think of it as a ship experience.

"We don't want your face buried in your mobile phone," he went on to explain. "We want to make sure that your primary experience is of the ship itself. This experience takes place in the most incredible setting. You're on this gorgeous ship that already has all of these immersive environments."

The philosophy behind the design of the game, according to Feder, has been "heads up" as much of the time as possible. "You may look at the phone for certain pieces," he said, "but in a lot of our interactives, all of the action is happening on the ship itself on a digital sign in front of you or as part of a prop, and your phone just becomes your controller. You don't even need to look at it. It's tracking your gestures, so you just tilt it or swirl it or sing into it and the microphone will pick up the sound. We're using your phone and all of its capabilities without you having to worry about it."

So what, exactly, is happening on your Disney Uncharted Adventure?

"Just as sailors and wayfinders have been looking at the stars for inspiration and guidance for centuries at sea," said Feder, "we looked to the stars for guidance and inspiration for our story, and, really, one star in particular, the Wishing Star. Everyone knows when you wish upon a star your dreams come true. Disney Uncharted Adventure flips that on its head and says, 'Okay then, what would happen if that Wishing Star were to go missing?'"

The experience begins with Captain Minnie and Captain Mickey taking you on a stargazing journey of the sky above the *Disney Wish*, but as so often happens in the best Disney stories, something goes awry.

The Wishing Star, the center of the Disney storytelling universe, gets knocked from the sky and shattered into pieces that land in various Disney story worlds, such as Never Land, Motunui, and New Orleans.

"It becomes your quest, your hero's journey," said Feder, "to go to each of these story worlds, interact with characters there, find that piece of the star, face off with some perilous foes who might want the star for their own nefarious purposes, and recover all those pieces to get the star back in the sky."

Naturally, there will be games and challenges, but guests don't have to be gamers to flourish.

"Our games are designed so that no matter how well you do, you're always succeeding," said Feder. "The role of our interactive games is to give you agency in the story. You are living the adventure.

"For instance," he continued, "we want you to help navigate Moana out of the reef. Even if you crash into every single shoal along the way, you've helped her. If you're great at it, great. You'll get to feel really good at the end and the puzzles will get harder and that's awesome. But if you're not great at it, that's okay, too. The reason we wanted you to do that is for you to take a role in the story, rather than sitting back."

When a potentially cataclysmic event occurs, guests are sent on a series of quests in which they encounter characters from their favorite Disney films, such as the Kakamora from *Moana* (top left and center left). Accomplishing certain tasks activates effects on the ship (bottom left). Guests start at home by creating their own avatar, who can earn items and objects during the course of game play (bottom right). Philip Gennotte tries out Disney Uncharted Adventure during a "playtest" of the technology (top right).

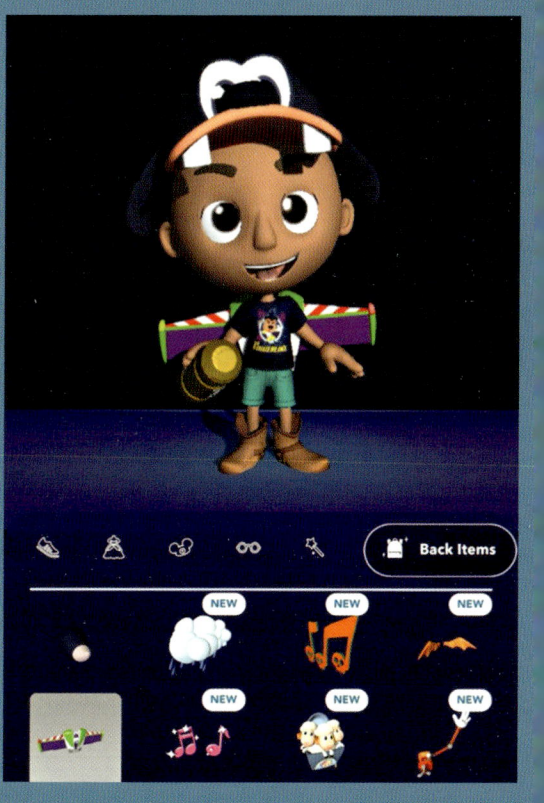

Captain Mickey and Captain Minnie are the hosts for Disney Uncharted Adventure, guiding guests through the quests and seeking to restore order when things inevitably go awry.

The best part, though, may be the last step, where everyone comes together to finish the job, so to speak.

"The culmination of the experience," said Feder, "is where everyone who has been on this adventure—and even those who have been hearing the buzz and just want to join in—will come together to participate collectively in putting those pieces of the star back together and getting that star back up in the sky, giving everyone that finale moment."

But with this being Disney, it's never that easy.

"Of course, things won't go entirely according to plan," said Feder, "because you realize that there's been somebody who's been chasing down the Wishing Star this entire time and pulling a lot of the strings behind all of the activity. And she's going to crash the party and try to steal the star for herself. [So] what started as a simple reassembly task will end up turning into an epic battle between good and evil as you fight over control of the star.

"It's a great opportunity for something to happen at a specific place and time," Feder added, "which a lot of interactives and gaming experiences miss because they're often experienced independently and individually. This is a chance for something to be experienced together, by the entire group."

"It's not the same as something you can plop into Magic Kingdom or plop into the theaters," said Feder. "Because of the way we designed it, taking advantage of its environment, Disney Uncharted Adventure is an interactive experience that can only take place on [this] Disney cruise ship."

Guests go on quests to as many as six story worlds, each of which features games and activities, such as making music with Hector in *Coco*, that get them closer to solving the mystery at the center of Disney Uncharted Adventure.

Davey Feder

Software Product Manager, Walt Disney Imagineering

Just about every summer when he was growing up in the Bay Area around San Francisco, Davey Feder, the lead designer on Disney Uncharted Adventure for the *Disney Wish*, went on a family vacation to Disneyland Park in Anaheim, California.

"My birthday is in summer," he said, "so it always made it extra special for me."

Maybe that's why he became the biggest Disney fan in the Feder family, far eclipsing the interest shown by his older sister and younger brother.

"I think Disney just fit with my own personality and passions," he said. "Disney's all about story and I loved a world where everything had a story and every story made sense."

It helped that Davey came of age at a great time for Disney movies. "I remember seeing *The Lion King* and *Aladdin* and all those films, and I was totally enchanted by all of them.

"Plus," he added, "I was a Disney Afternoon kid with *Chip n' Dale Rescue Rangers* and all that great Disney stuff in the nineties."

Then, in middle school, Davey learned what Walt Disney Imagineering was.

"I saw these TV specials that told the story of how the parks get built and the people who do it. That was what first hooked me."

A few years later, he graduated from Stanford University with a degree in Science, Technology and Society, a major focused on "human-centered" design.

"I thought would be a good path to Imagineering," he said.

Turns out, Davey was right, although the trail was a little twisted and filled with more than its share of pitfalls and distractions.

First, he was hired by a gaming company that had just been bought by Disney. Then came a stint at Lucasfilm, which had also been purchased by Disney.

But it still wasn't Imagineering. Finally, in 2019, he got "the call."

"'Are you interested?'" Davey said he was asked. "I could hardly contain my enthusiasm. Imagineering was the thing that brought me to Disney in the first place and now I finally had a chance to go there myself.

"If eleven-year-old me could see me now, this is always what he wanted to do. I made it to this place I always wanted to be and it's a really cool feeling to see a childhood dream fulfilled like this."

James Willoughby Director, Hotel Operations, Special Projects, Disney Cruise Line

After graduating with a degree in hotel business management in his native England, and following stints at the Walt Disney World Resort and in hotels in Hong Kong, James Willoughby found himself in Australia, gaining experience working at Peninsula Hotels and the Sydney Opera House.

"I was sitting in Sydney Harbor, near the end of my time in Australia, and I saw two cruise ships sailing in," he said. "I just sat there by myself, staring at them, and I thought, 'It would be kind of interesting to look at that industry.' I flew back home to London and applied for several leadership roles in the cruise industry," he said, continuing the story. "Thanks in part to my experience at Walt Disney World, I was offered a corporate trainer position for Carnival Corporation based in Miami."

That experience was intense, but brief ("I actually went on about thirteen ships in nine months," he said), because Disney Cruise Line had just launched the *Disney Magic* and a former contact reached out to him to join the Disney team. It was the first of many roles he would have in the industry with Disney as well as other companies. But he always seemed to find his way back to Disney Cruise Line, returning in operational roles, as well as playing a key role in the construction and delivery of the *Disney Dream*, *Disney Fantasy*, and, now, the *Disney Wish*.

Shelley Gold-Witiak Director, Hotel Operations—Lodging, Disney Cruise Line

Growing up in the middle of England, in Wolverhampton, Shelley Gold-Witiak could hardly imagine visiting a Disney park or sailing on a cruise ship, let alone one day working at and/or on one.

"Our vacations were never abroad, always at the coast in the south of England," she said. She did, however, embark on a career in the cruise industry, working for several different cruise lines for more than two decades when, in 2004, Disney Cruise Line came calling.

That ultimately was a turning point for me," she said. "I've never looked back, and I always say it was the best decision, not only for me, but also my family."

In the eighteen years since, Shelley has helped launch the *Disney Dream* and *Disney Fantasy*, and played a key role in introducing concierge lounges across the fleet.

Today, she oversees a team focused on the staterooms and concierge experience for the *Disney Wish*.

Now Shelley lives in Celebration, Florida, with her husband, Bill, whom she met on board while the two worked together at Royal Caribbean. They have two grown daughters.

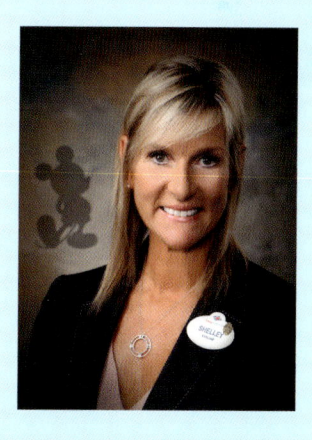

The majestic *Disney Wish*, just before floating out of the construction hall at Meyer Werft.

A Wish Fulfilled

When the *Disney Wish* emerged from Hall 6 at Meyer Werft on that early February day in 2022, it *looked* finished—its exterior fresh, shiny, and pristine. But the ship still had a long journey ahead before reaching Port Canaveral, Florida, ahead of its maiden voyage in summer 2022. For one thing, there was still a lot to do *inside* the ship, from completing the interior spaces to continuing the testing of all the marine, technical, and safety systems.

In late March, after being docked alongside a pier at its Meyer Werft's home in Papenburg, Germany, for all of that additional work and testing, it was time for the *Disney Wish* to say goodbye to the shipyard—literally.

In a Meyer Werft tradition that comes with the departure of every ship, "Time to Say Goodbye," sung by Andrea Botticelli, is played on large speakers on the promenade deck as the ship starts her conveyance down the river Ems.

"It's an emotional moment for the yard," said Philip Gennotte. "It's like saying goodbye to your baby. For us, we're taking the ship. It's going with us. But for the yard workers, this is it. They will move on to the next ship and likely never see her again."

Following the conveyance down the river Ems, the *Disney Wish* made a short stop in Eemshaven, Netherlands, for a few days of inspections before embarking on a sea trial, the first in a number of tests that assessed the ship's sea worthiness.

From there, it was off to Bremerhaven, Germany, where the *Disney Wish* bunkered her first full supply of LNG fuel and prepared for her main sea trial.

Finally, in the spring, came another huge moment . . . a threshold for the *Wish*: the handover of the ship from Meyer Werft to Disney Cruise Line.

"The last few days before that were very hectic," said Gennotte. "Both sides, the yard and Disney, had been working nonstop, 24/7, to get all the paperwork, all the documentation to make sure everything was in place for the final protocol of delivery. It was massive loads of paper and all the attachments [that go with it] that we had to get ready."

In a sense, it was like making the final arrangements to buy a house, making sure the inspections had been done, the title was clear—and the mortgage was approved.

"I remember on the *Disney Fantasy*," said Gennotte, who at that time was working for Meyer Werft, "we were sitting in the Concierge Lounge. We had a table set up, nicely done, we're all in our suits and ties, and we have the final paperwork ready to be signed by Disney and Meyer Werft.

"But it was the craziest thing," he continued, laughing. "We were all sitting there looking at each other, waiting for the signal from the finance teams. Finally, they said, 'The money's there, you can sign.'"

After the papers were signed aboard the *Disney Wish*, there came yet another emotional moment. (Disney ships are seemingly built for an endless parade of emotional moments—just recall Mo Landry's comments about Deck 14 and the ties guests form with one another when on a Disney Cruise.)

"There was a ceremony on the top deck," said Gennotte, "where the German and Meyer Werft flags were exchanged for the Disney Cruise Line flag.

"We had music and of course the picture moment, and then the official handover of the book that says, nicely, *Disney Wish* on it," he continued. "And then the yard captain, who had operated the ship until that moment, handed [it] over to the Disney Cruise Line captain, Marco Nogara. It was literally like handing over the keys to a car, only it's a big ship."

With the *Disney Wish* now in full possession of Disney Cruise Line, the ship set off for its transatlantic crossing, still very much in pre-opening mode, with Imagineers still scampering around providing the final touch-ups to the spaces and the last adjustments to show and technical systems, while crew members and performers went through training and rehearsals.

As our story of the making of this magnificent ship comes to an end, though, the *Disney Wish* is just beginning its journey, but not before we hear a few parting words from a couple of those who worked so long and hard to bring her to life.

"The scale and scope of a ship like this is significantly more than a resort," said David Atwood, principal project manager, Walt Disney Imagineering. "We hear people say that cruise ships are like floating hotels. That parallel is

"You can dream, create, design, and build the most wonderful place in the world," Walt Disney famously said, "but it requires people to make the dream a reality." (Top left) Captain Marco Nogara, Steve Read, Giacomo Panicucci, Staff Captain Marko Paas, Philip Gennotte, Adriano Patteri, Gareth Hamblin, Bert Swets, and Zvonimir Vidak. (Top right) James Willoughby, Mo Landry, and Meghan Moore. (Bottom left) Lauren Fong. (Bottom right) Patrick Burnett, Michael Weyand, and Doug Larsen with the *Disney Wish* horns.

Counterclockwise from far left: The in-field AquaMouse team included (front, left to right) Lauren Fong, Pete Leathers, Yara-Lisa Gewetzki, and Manfred Erb; (second row) Alison Freedman, Matthias Lampe, Maura King, and Jan Rothhaupt; (back row) Mike Sigmund, Ulrich Grohman, Jay Cardinali, and Stephan Spiller. Lauren Fong welcomes the gigablock back into the construction hall at Meyer Werft. Grayson Bilot watches a video on the *Disney Wish*. A wall signed by those who worked on the ship (it's near Enchanté, but not publicly visible).

certainly understandable because you're staying in one property and you're staying in guest rooms . . . and you have certain amenities. But the scale and scope of a ship is so much greater. You don't see too many hotels with Broadway-style theaters, three 700-seat restaurants, and so many bars, shops, and different venues."

Added Bob Tracht, "I don't know how we did it, but we have. We've had so many seemingly impossible projects that we look at and go, 'We're never going to get this done.' And yet, we do. We have incredibly dedicated people around the world and, even in a pandemic, we were able to make some modifications to our process and keep things moving."

Mo Landry was particularly pleased with how the ship turned out, given the observations some people made after the ship's initial unveiling at a virtual event on April 29, 2021, more than a year before its official launch.

"After the 'Reveal,'" she said, "there were a few comments about how it's almost the same as the previous ships, and I just kept thinking to myself, *Oh, just wait. You need to see. We're only showing bits and pieces*. Sure enough, once they truly understood what the ship was really all about, they were blown away. It was amazing."

There was also no shortage of Imagineers who were looking to take a break after their years of exhaustive and exhausting work on the *Disney Wish*.

One, David Atwood, was looking forward to a little peace and solitude, "kicking back and having some cocktails in the Quiet Cove area."

Walt Disney Imagineering assistant project manager Lauren Fong was "ready to christen the piano in Nightingale's."

And Claire Weiss, who "hopes that our guests feel a lot of the magic and the heart that we have poured into this project," was also looking forward to her moment of rest.

"I particularly liked the part where the ship was finished and I was having a piña colada on the banks of the river Ems while it sailed past," said the Walt Disney Imagineering creative director looking back at the *Wish*'s creation and launch.

Danny Handke, upon reflection, seemed all in on a longer break once the *Wish* was in operation. "The *Disney Wish* has the perfect balance of theme and Disney storytelling," he said. "For me, it checks all the boxes creatively, from Disney and Pixar to *Star Wars* and Marvel. I felt like I could retire after this."

But that's not likely.

As Walt Disney said on the tenth anniversary of Disneyland in 1965, "I just want to leave you with this thought, that it's just been sort of a dress rehearsal. . . . So if any of you start resting on your laurels, I mean just forget it, because . . . we are just getting started."

And that's certainly true in this case because the *Disney Wish* is just the first of three ships to be delivered by this team between now and 2025.

Still to come are the *Disney* . . .

Alyssa Markfort

Associate Show Manager, Walt Disney Imagineering

"A lot of Imagineers, it was always their dream to come work for Walt Disney Imagineering," said Alyssa Markfort," but I kind of fell into it. It wasn't something that was even on my radar to come and work for Disney, certainly not Imagineering."

It wasn't because she didn't like Disney.

"I grew up with Disney Channel and the resurgence of Disney animated films," she explained. "*The Little Mermaid* was my absolute favorite movie. I saw it when it was rereleased in theaters and then I got it as a gift, a VHS tape—I think my parents still have it—that is completely tattered and just totally worn out, but we just never had the opportunity to go to the parks."

Part of the reason is that her family moved—a lot.

"I was born and grew up mostly in St. Cloud, Minnesota," said Alyssa, "but we also lived in Arizona and Virginia."

The family finally settled in Virginia Beach, Virginia, where her love of theater and performing blossomed, first in high school and then across the state at Virginia Tech, where she majored in stage and production management.

College is also where she finally got her first taste of the Disney parks.

"I finally got to go during my sophomore year," she said. "Everybody laughed, 'You're going to Disney for spring break?' And I was like, 'Absolutely, yes.'"

Still, Alyssa never thought of Disney as a career.

"As I was finishing up my degree, I honestly didn't know what I wanted to do with my life," she said.

And that's when she discovered that Disney might be the place for her, after all.

"Walt Disney Imagineering—Florida was looking for an intern, specifically someone with a stage management background," Alyssa said. "I couldn't believe it. Really? I applied, got an interview, and a couple weeks later I was hired. I had to move to Florida, but that was something I gladly did."

She was so excited about her new job that, a few months after joining Imagineering, she tweeted, "If Disney would send me to Europe for work as long as they want my life would be complete."

She got her wish when she started working on the *Disney Wish*, which included a move to Germany in 2019.

Her job has included managing the production of the sculpted characters found throughout the ship, including Cinderella in the Grand Hall and Rapunzel on the stern.

(Top left) Jan Krefting and and Anton Erasmus of design partners YSA Design are pleased with the progress.

(Center left) Kyle Bilot, Meghan Moore, Mo Landry, Alyssa Markfort, and Bob Girardi in front of the *Disney Wish* gigablock.

(Top right) Tim Hall takes a moment to survey all the work being done.

(Bottom left) Tom Hanssen in the middle of an inspection.

(Bottom right) Kristen Zeigler, Karyn Poore, Leonard Lopez, and Katie Eastman send a truckload of props on their way to Papenburg.

. . . well, at this point, let's just call them Disney Ship 6 and Disney Ship 7.

For the very last word, though, let's turn to Laura Cabo. "The *Disney Wish*, for me," she said, "was a passion project where I could bring my whole self with my dad, who was in the military nearly his entire life and was always around large ships. I have such an emotional connection to this project, working on a technically advanced, wonder of the world like this cruise ship. I know that my father, who passed away a decade ago, would be so proud.

"Most importantly for me is that it seems like yesterday that we scribbled three words into a notebook and now they are our ships. I am in awe of how these seeds of an idea were taken to heart by hundreds, no, thousands of dedicated and talented people. Together we have created the most magical ships to ever sail the seas.

Laura Cabo (left) takes a moment to thank her boss for the opportunity of leading the creative team on the *Disney Wish*. (Actually, she's at the "Mickey: The True Original Exhibit" in New York City.)

This Page: (top left) Jonathan Frontado, Evily Peros, Tiffany Deitz, and Mary Precourt. (Bottom right) Sebastian Schwartz-Weihs of Meyer Werft and Ingo Eilers celebrate another success. (Bottom left) Nicola Bruiglia. (Top right) After a long day, Markus Rauna catches a nap in the still-under-assembly crew quarters.

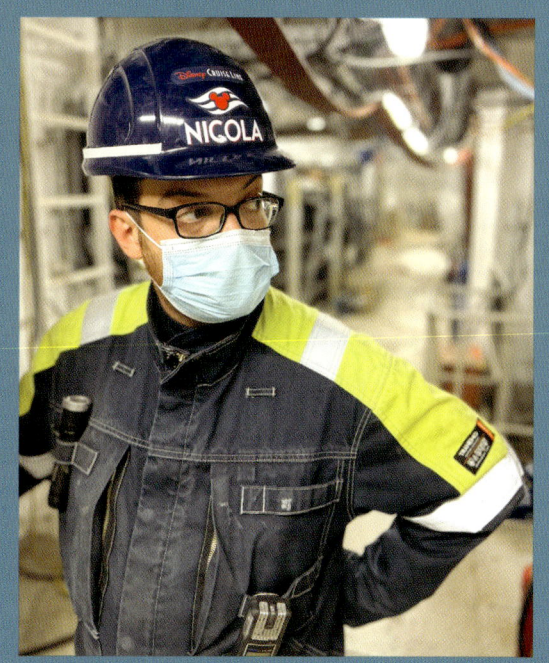

Disney WISH
NASSAU

Making a Disney WISH
Behind the Dreams of Disney's Newest Ship